The Institute of Biology's
Studies in Biology no. 29

Size and
Shape

by R. McNeill Alexander
Ph.D., D.Sc., F.I.Biol.
Professor of Zoology, University of Leeds

Edward Arnold

First published 1971
by Edward Arnold (Publishers) Limited,
25 Hill Street,
London, W1X 8LL

Reprinted 1973
Reprinted 1975

Boards edition ISBN: 0 7131 2333 8
Paper edition   ISBN: 0 7131 2334 6

Printed in Great Britain by
The Camelot Press Ltd, Southampton

General Preface to the Series

It is no longer possible for one textbook to cover the whole field of Biology and to remain sufficiently up to date. At the same time students at school, and indeed those in their first year at universities, must be contemporary in their biological outlook and know where the most important developments are taking place.

The Biological Education Committee, set up jointly by the Royal Society and the Institute of Biology, is sponsoring, therefore, the production of a series of booklets dealing with limited biological topics in which recent progress has been most rapid and important.

A feature of the series is that the booklets indicate as clearly as possible the methods that have been employed in elucidating the problems with which they deal. Wherever appropriate there are suggestions for practical work for the student. To ensure that each booklet is kept up to date, comments and questions about the contents may be sent to the author or the Institute.

1971

Institute of Biology
41 Queen's Gate
London, S.W.7

Preface

This booklet reflects a growing realization among biologists that physics and engineering theory can be used to elucidate the anatomy of plants and animals. It includes many rough (and very simple) calculations which are designed to show that the explanations which are offered are plausible. There is room in it to consider only a few of the most striking features of some of the major groups of plants and animals. I have dealt mainly with external features and I have written less about plants than about animals for the sound but regrettable reason that I know less about them.

I have obtained some of the information for this booklet by pestering my colleagues in the University of Leeds. I am very grateful to them.

Leeds, 1971   R.McN.A.

Contents

Size and Proportions

1.1 Allometry

A constant theme of this book is that changes of size make changes of shape and proportions necessary. Not only are the proportions of a chick quite different from those of an adult hen, but the proportions of a small species of bird are quite different from those of a large one (Fig. 1–1). 'Allometry' is a term used to describe relationships between dimensions of organisms; that is, to describe the effects of size on their proportions. The term refers both to comparisons between specimens of different size or age but of the same species, and to comparisons between different species (GOULD, 1966). This book is mainly concerned with comparisons between species.

(a) **(b)**

Fig. 1–1 Outlines traced from photographs of (**a**) a fairly small bird, the Ringed plover (*Charadrius hiaticula*, photograph by Miss I. Werth) and (**b**) a very large bird, the Kori bustard, *Ardeotis kori*. (Photograph by Dr. J. Brust) Not to the same scale.

Data on allometry can be presented in various ways. Pictorial comparisons, such as Fig. 1–1, may be interesting, but they are often less informative than quantitative comparisons. A quantitative comparison may be made simply by tabulating the data (which are apt to be indigestible in this form) or by presenting them as a graph or as a mathematical equation. We will consider how graphs can best be used and how equations can be obtained, using for our examples data about birds.

Figure 1–2 is a graph of the weight of the flight muscles against the total weight of the body for a selection of birds. The points are scattered reasonably evenly on either side of a straight line which passes through the origin. This indicates a tendency for flight muscle weight to be proportional to body weight. However, the points are scattered and show, for

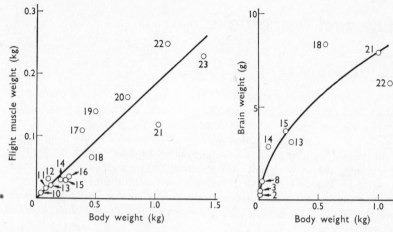

Fig. 1–2 Graph of the total weight of the flight muscles against body weight, for some of the birds listed in the table below. (Data from GREENE-WALT (1962))

Fig. 1–3 Graph of brain weight against body weight for some of the birds listed in the table below. (Data from VAN DER KLAAUW (1948))

Species of bird for which data are given in Figs. 1–2 to 1–6

1. Hummingbird, *Eupherusa eximia*
2. Goldcrest, *Regulus regulus*
3. Wren, *Troglodytes troglodytes*
4. Blue tit, *Parus caeruleus*
5. Robin, *Erythacus rubecula*
6. Chaffinch, *Fringilla caelebs*
7. Skylark, *Alauda arvensis*
8. House sparrow, *Passer domestica*
9. Swift, *Apus apus*
10. Sanderling, *Calidris leucophaea*
11. Starling, *Sturnus vulgaris*
12. Snipe, *Gallinago gallinago*
13. Common tern, *Sterna hirundo*
14. Magpie, *Pica pica*
15. Kestrel, *Falco tinnunculus*
16. Barn owl, *Tyto albo*
17. Partridge, *Perdix perdix*
18. Carrion crow, *Corvus corone*
19. Wood pigeon, *Columba palumbus*
20. Curlew, *Numenius arquatus*
21. Buzzard, *Buteo buteo*
22. Mallard, *Anas platyrhynchus*
23. Heron, *Ardea cinerea*
24. Gannet, *Sula bassana*
25. Golden eagle, *Aquila chrysatus*
26. Swan, *Cygnus cygnus* or *C. olor*
27. Vulture, *Gyps fulvus*
28. Albatross, *Diomedea exulans*
29. Great bustard, *Otis tarda*

instance, that the buzzard (**21**) has small flight muscles for its size while the wood pigeon (**19**) has large ones. These particular deviations from average proportions are not very surprising since the buzzard spends a lot of time soaring in thermals and makes rather little use of flapping flight, while the pigeons are strong fliers. The flight muscles make up a variable proportion of the body weight but there does not seem to be any systematic change of this proportion with size.

Figure 1–2 is not as useful a graph as might be wished. The data for small birds are squashed into the bottom left-hand corner so as to make it impossible to show clearly that a 3 g hummingbird has a higher proportion of flight muscle in the body (34 per cent) than any of the other species, and even so it has been impossible to fit in birds larger than a heron.

Figure 1–3 is another straightforward graph, this time of brain weight weight. Brain weight does not seem to be proportional to body .5 per cent of body weight in a 6 g goldcrest, 1.1 per cent in a 1 only 0.35 per cent in a 5 kg eagle (too heavy a bird to be the graph). There is a general tendency for brain weight to be aller proportion of body weight as size increases, which largely explains why small birds have relatively larger heads than large ones (Fig. 1–1). This is not surprising: there is no reason to think that the complexity of the tasks the brain has to perform should increase in proportion to the weight of the body. No straight line could represent the tendency of the points as well as the smooth curve which has been drawn through them and which, incidentally, draws attention to the fact that the crow (**18**) has a large brain for its size and the mallard (**22**) a small one. It would be possible by trial and error to find an equation which fitted the curve reasonably well, but there is no really quick method for finding the equation from the graph plotted in this way.

Figure 1–4 shows the same data as Figs. 1–2 and 1–3, re-plotted on logarithmic coordinates. Special graph paper has been used which saves the trouble of using tables to convert the data to logarithms, but the effect is exactly the same as if the logarithms of flight muscle and brain weights had been plotted against the logarithm of body weight. This method of plotting has made it possible to fit in data for much heavier birds than could be included conveniently in the other graphs, without squashing up the data for small birds. Indeed, the points for small birds are spaced out much more than in the other graphs and it is much easier to see whether a small bird has larger or smaller muscles and brain than is usual for its size. Another effect of this method of plotting is that it has presented the brain weight data so that a straight line can be drawn through the points as satisfactorily as for the muscle graph.

We will now use Fig. 1–4 to obtain equations relating flight muscle and brain weights to body weight. Consider two variables x and y related by the equation

$$y = kx^\alpha \qquad\qquad (1.1)$$

where k and α are constants. By taking logarithms we find that

$$\log y = \log k + \alpha \log x \qquad\qquad (1.1a)$$

so that a graph of $\log y$ against $\log x$ (or a graph of y against x plotted on logarithmic coordinates) must be a straight line of gradient α. Conversely, if a graph on logarithmic coordinates of y against x turns out to be a straight line, x and y must be related by an equation of the same form as **1.1**. Such

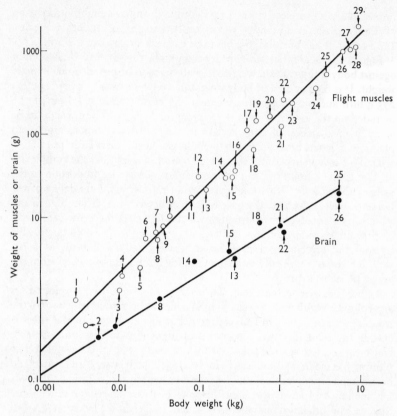

Fig. 1–4 Graphs on logarithmic coordinates of flight muscle weight and of brain weight against body weight, for birds listed in the table on p. 2. (Data largely the same as for Figs. **1–2** and **1–3**, but additional data included from the same sources)

relationships are found very frequently indeed in investigations of allometry, and Fig. 1–4 provides us with two examples.

The exponent α in equation **1.1** is the gradient of the graph of $\log y$ against $\log x$. There is danger of confusion if graph paper with logarithmic coordinates is used, as in Fig. 1–4. The values of $\log x$ and $\log y$ must be used in calculating the gradient, not x and y themselves. This gradient for the flight muscle line is 0.96, so

$$\text{flight muscle weight} = k \, (\text{body weight})^{0.96}$$

Both weights are given in kilograms. When body weight is 1 kg flight muscle weight is 0.18 kg so $k = 0.18$. Our equation is thus

$$\text{flight muscle weight in kg} = 0.18 \, (\text{body weight in kg})^{0.96} \qquad (\textbf{1.2})$$

Similarly for the other line

$$\text{brain weight in g} = 8.2 \, (\text{body weight in kg})^{0.60} \qquad (1.3)$$

It must be emphasized that these equations cannot tell us the exact weights of the flight muscles and brain of, for instance, a bird weighing 0.56 kg. They only give us average expected values for the weights of muscle and brain.

The lines in most of the graphs in this book have been fitted to the points by eye, and there would have been very little risk of error if the flight muscle line (Fig. 1-4) had been so fitted: the points are scattered so little that it is pretty obvious where the line should be drawn. However, this particular line is a regression line fitted by a statistical method which is described by BAILEY (1959). The analysis was of course applied to the logarithms of the weights of muscle and body, not to the weights themselves. It appears from the analysis that the gradient of 0.96 is probably just significantly different from 1.00 (probability $\simeq 0.05$), but there is no serious conflict here with our conclusion from Fig. 1-2, that flight muscle weight tends to be proportional to body weight.

Brain weight has been found to be more or less proportional to (body weight)$^{0.6}$ in many different groups of vertebrates as well as in birds, though the constant of proportionality k varies from group to group (VAN DER KLAAUW, 1948).

1.2 Length, area and volume

Imagine two organisms of which the smaller is an exact scale model of the larger. In the language of allometry, these organisms are isometric in all respects. If the larger is n times as long as the smaller any organ in it is n times as long, n times as broad, and n times as thick as the corresponding organ in the smaller one. The organ must therefore have n^2 times the surface area and n^3 times the volume of the corresponding organ in the smaller organism. If it is constructed of the same materials in both, its weight in the larger organism is n^3 times its weight in the smaller. Conversely, lengths of corresponding parts in isometric organisms would be proportional to (body weight)$^{0.33}$, areas to (body weight)$^{0.67}$, and volumes or weights to (body weight)$^{1.00}$.

The data of the previous section show us that birds tend to be almost isometric in the relationship between flight muscle weight and body weight, but not in the relationship between brain weight and body weight. Now look at wing span and body weight (Fig. 1-5). If the relationship were an isometric one wing span would be proportional to (body weight)$^{0.33}$, but the graph indicates that it tends to be proportional to (body weight)$^{0.39}$. There is a small but distinct departure from isometry which will be referred to again in Chapter 4. Similarly wing areas would be proportional to (body weight)$^{0.67}$ if birds were isometric, but are actually more nearly proportional to (body weight)$^{0.76}$.

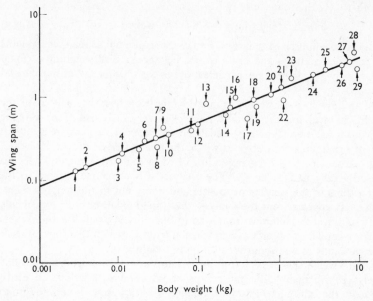

Fig. 1–5 Graph on logarithmic coordinates of wing span against body weight for birds listed in the table on p. 2. (Data from GREENEWALT (1962))

A botanical example will serve as a reminder that this sort of analysis can be applied to plants as well as to animals, and to specimens of varied size of the same species as well as to specimens of different species. WHITTAKER and WOODWELL (1968) made many measurements on trees and shrubs, including the diameter of the base of the stem and (by estimation from samples) the total area of the leaves. They found that leaf area was proportional to $(diameter)^{1.5}$ in one species, to $(diameter)^{2.2}$ in others and to intermediate powers in the rest. If the relationship were isometric leaf area would of course be proportional to $(diameter)^2$.

1.3 Size, time and power

So far we have used graphs on logarithmic coordinates to show relationships between the body weights of birds and the weights and dimensions of various parts of the body. Figure 1–6, also on logarithmic coordinates, shows how the frequency of the wing beat depends on body size. Large birds make fewer beats per second than small ones. The points for birds other than hummingbirds are scattered fairly evenly on either side of a straight line of gradient −0.26, showing that wing beat frequency tends to be proportional to $(body\ weight)^{-0.26}$. The figures for hummingbirds seem to refer to hovering and so are not strictly comparable with the figures for normal forward flight of other birds. Pigeons beat their wings faster when they hover (which they can only do for a few moments) than in forward flight (PENNYCUICK, 1968b).

The frequencies of other repetitive movements decrease similarly as body weight increases. The frequency of the heart beat of birds is about proportional to (body weight)$^{-0.2}$, whether the frequency at rest or the maximum frequency in activity is considered (BERGER *et al.*, 1970). Much the same is true of other groups of vertebrates. The frequency of the breathing movements of mammals seems also to be about proportional to (body weight)$^{-0.2}$. A running mouse takes more steps to the minute than a horse, a dog or even a rat, and the stepping rate at the speed at which these animals change from a trot to a gallop is proportional to (body weight)$^{-0.14}$ (HEGLAND, TAYLOR and MCMAHON, 1974).

The general rule that large animals need more time for each movement than small ones can be explained by considering the work done by a muscle when it contracts (HILL, 1950; SMITH, 1968). The force it exerts can be expected to be proportional to its cross-sectional area. The distance it can move its point of insertion is proportional to its length. The work done in a single contraction is obtained by multiplying the force by the distance, so it must be proportional to (cross-sectional area of muscle) × (length of muscle); in other words, it must be proportional to the volume of the muscle.

We will now imagine a series of animals of differing weight which are isometric in all respects. If the body weight of one of these animals is W, the volume of a particular muscle which we will consider is cW, where c has

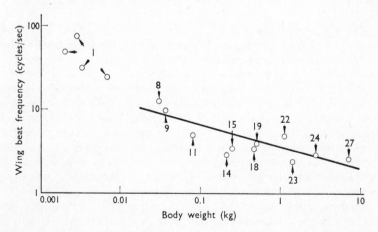

Fig. 1–6 Graph on logarithmic coordinates of wing beat frequency against body weight, for birds listed in the table on p. 2 and for some other species of hummingbird. (Data from GREENEWALT (1962))

the same value for the whole series of animals. If a unit volume of muscle can do work c' (which we will take to be constant) in a single contraction, this muscle can do work $cc'W$. Suppose this work is used to give velocity V to a portion of the body of mass $c''W$, where c'' is another constant for the series of animals. The work done by the muscle equals the kinetic energy gained by this part of the body and

$$cc'W = \tfrac{1}{2}c''WV^2$$
$$V = (2cc'/c'')^{\frac{1}{2}} \tag{1.4}$$

Thus V is a constant, independent of W. Isometric animals can accelerate corresponding parts of their bodies to the same velocity, irrespective of body weight.

Let us apply this conclusion to the feet of animals. Similar animals should be able to accelerate their feet to the same velocity, and so to run at the same speed, irrespective of their size. This explains why there is so little difference between the top speeds of whippets (15 m/s) and greyhounds (16 m/s, HILL, 1950) although greyhounds have legs about 50 per cent longer than those of whippets. It also explains why large animals take fewer steps per minute than small ones: the distance the foot can be moved, relative to the body, is proportional to the length of the limb or to $W^{0.33}$, so if it moves this distance at the same velocity in animals of different size the time required is proportional to $W^{0.33}$. Hence the number of steps taken per minute should be proportional to $W^{-0.33}$. If the work done in each contraction of the muscles is, as we have assumed, proportional to W and the frequency of contraction to $W^{-0.33}$, the power output of the muscles (i.e. the rate at which they do work) must be proportional to $W^{0.67}$

This argument depends on the assumption that most of the work being done by the muscle is being used to accelerate the limb. This assumption may not be sound even for the limbs of mammals running on level ground. The argument does not at first sight seem applicable to the wing beat of birds since most of the work is done against air resistance and only a little of the work is used to accelerate the wings. However, this acceleration must occur at the beginning of the downstroke, when most of the work being done may be needed for it (PENNYCUICK, 1968). The low beat frequencies of large birds may thus be at least partly explained in the same sort of way as the low stepping frequencies of large running animals. If so, the effect must be accentuated by the fact that large birds are not isometric with small ones but have relatively larger wings (Fig. 1–5).

Attempts have been made to explain the relationship between heart rate and body size in the same sort of way. Such explanations take the heart as the muscle of volume cW referred to in equation 1.4, and the blood ejected from the heart at a single contraction as the part of the body of mass $c''W$ which is given velocity V. These explanations are unconvincing as it can be calculated from the velocity and pressure of blood leaving the heart that only 1–2 per cent of the work done by the human heart is used to accelerate the blood. (Most of the work is done against viscosity, in driving the blood through the finer blood vessels.)

1.4 Growth with shape unchanged

The hard skeletons of many animals and the wood of trees consist largely of non-living material which can be added to, but cannot grow. The shell of a crab cannot grow but is cast off and replaced by a new soft one, developed inside it, which is stretched to a larger size before it is hardened. Bone cannot grow but bones are enlarged by adding new bone outside the old, and if the bone has a cavity in it this is enlarged by removing some of the old bone. Trees grow by adding new wood outside the old, and a tree could be pared down so that all that remained was the wood of the sapling from which it grew. Similarly snail shells grow by the addition of material, leaving the shell of the young snail as the small whorls at the apex of the adult shell. However, while the tree changes its shape as it grows the snail shell does not (THOMPSON, 1942). It is a striking exception to the general tendency for animals to change their proportions as they grow. A hemispherical shell could not be added to without changing its shape, nor could a cylindrical one be added to without changing its proportions. However, a conical shell like that of a limpet can be added to without change of shape or proportions (Fig. 1–7(a)). The same is true of the shape shown in Fig. 1–7(b) which resembles the shells of many molluscs and is in effect a cone coiled into a spiral. The spiral seen in the side view is the curve known as the logarithmic spiral, which has the special property that the radius is multiplied by the same factor at every

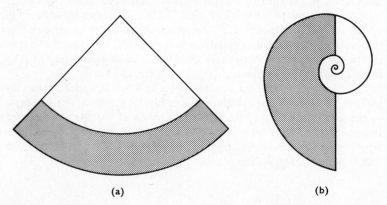

(a)   (b)

Fig. 1–7 Diagrams of a conical and a spiral mollusc shell. Note that in each case addition of the tinted region to the original untinted shell increases the size of the shell without altering its shape.

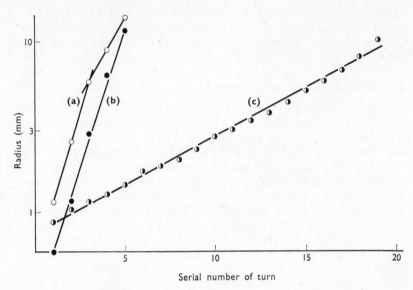

Fig. 1–8   Graphs of radius (on a logarithmic scale) against number of turns for the shells illustrated in Plate 1.

turn, that is, the radius r of the spiral is related to the number of turns n by the equation

$$\log r = kn \qquad\qquad (1.5)$$

where k is a constant, so a graph of $\log r$ against n is a straight line. Adding more turns to the spiral produces a curve indistinguishable from an enlarged photograph of the original spiral. Figure 1–8 shows how closely the outlines of the whorls of two of the mollusc shells of Plate 1 correspond in one case (b) to a logarithmic spiral and in the other (c) to a logarithmic spiral drawn out into a cone. Both shells grew with virtually no change of shape but shell (a) shows a sudden change of k at the third turn.

Curves which approximate quite closely to logarithmic spirals can also be found by cutting through the shells of bivalve molluscs, at right angles to the axis of the hinge joining the two halves of the shell. Such shells increase in size with little change of shape. The spirals are much less tightly coiled than those of the shells illustrated in Plate 1, which means that k (equation 1.5) is larger.

Metabolism

2.1 Metabolic rates

The energy on which the life of an animal depends is chemical energy bound up in the food which it eats. The animal releases this energy for use by the processes of chemical breakdown which are known collectively as metabolism, and the rate at which energy is used is called the metabolic rate. Metabolism may be either aerobic, requiring a supply of oxygen, or anaerobic, not requiring oxygen. Of these, aerobic metabolism yields far more energy from a given quantity of food. A monosaccharide unit in a glycogen molecule can be converted by aerobic metabolism to carbon dioxide and water, yielding 39 of the energy-rich phosphate bonds which are the immediate source of energy for many biological processes. Anaerobic metabolism of the same unit would convert it to a fatty acid such as lactic acid and yield only 3 energy-rich bonds. Most animals rely on aerobic metabolism, so their tissues must be kept supplied with adequate quantities of oxygen. The major part of this chapter is about the methods animals use to get oxygen to their tissues, and about how these limit their size or affect their shape.

Before we can discuss these problems we must know how much oxygen is needed. Some data are presented in Fig. 2-1 and, in a different form, in Fig. 2-2. As one would expect, animals use oxygen faster when they are active than when they are resting. Active animals, like insects, use oxygen faster than sluggish worms of similar weight, even when both are resting. The warm-blooded mammals and birds use oxygen faster than cold-blooded animals of similar size. Within each major group of animals oxygen consumption increases with body size but not (except perhaps in insects) in direct proportion to body weight. It is generally about proportional to $(\text{weight})^{0.75}$. This means of course that oxygen consumption per unit weight generally decreases as body size increases, and is about proportional to $(\text{weight})^{-0.25}$ (Fig. 2-2).

Figure 2-2 gives oxygen consumption as cm^3/g body weight hr. Some calculations which follow involve oxygen consumption as cm^3/cm^3 body volume hr. The animals we will be considering have densities close to 1 g/cm^3 and so it will be possible to assume, in comparing the data of Fig. 2-2 with the data of the calculations, that if the oxygen consumption is X cm^3/g hr it is also X cm^3/cm^3 hr.

2.2 Diffusion

Some animals get their oxygen direct from the atmosphere. Others, which live in water, use the oxygen dissolved in the water. Different

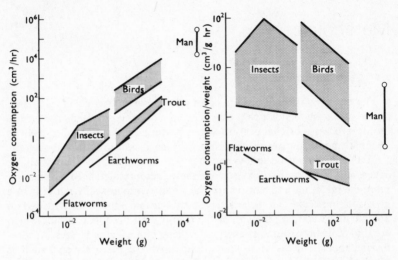

Fig. 2–1 (*left*)   The rate of consumption of oxygen against body weight for triclad flatworms (WHITNEY, 1942); tropical earthworms (LAVERACK, 1963); insects ranging in size from fruit flies to locusts (KROGH and WEIS-FOGH, 1951, and papers cited by them); the brook trout *Salvelinus fontinalis* (BROWN, 1957); birds ranging in size from hummingbirds to ducks (BERGER *et al.*, 1970); and man (PASSMORE and DURNIN, 1955). Where a range of oxygen consumptions is indicated for animals of the same type and weight the bottom of the range is the consumption at rest and the top is the maximum consumption during activity (in the cases of insects and birds, during flight).

Fig. 2–2 (*right*)   The data of Fig. 2–1 replotted to show oxygen consumption per unit body weight.

animals get the oxygen to their tissues in different ways. Some rely on diffusion through the superficial tissues to the deeper ones. Others have blood which circulates around the body collecting oxygen at the surface of the body or at gills or lungs and delivering it to the tissues where it is needed, but here again diffusion is involved: the oxygen must diffuse first into the blood, and then from the blood into the cells which need it. Yet others have air-filled tracheae through which oxygen diffuses to the tissues. Before we can go further we must understand the process of diffusion.

Diffusion of gases depends on gradients of partial pressure. The partial pressure of a gas in a mixture of gases is the pressure it would exert if it alone occupied the whole volume of the mixture. Thus air contains about 20 per cent oxygen by volume, so the partial pressure of oxygen in air at atmospheric pressure is 0.2 atmosphere. The partial pressure of a gas in solution is the partial pressure it would have in a mixture of gases which

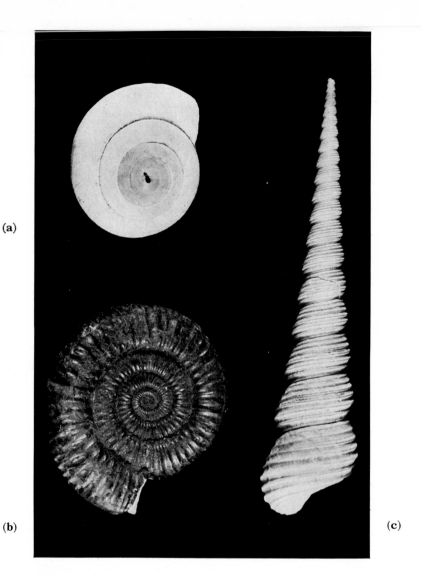

(a)

(b)

(c)

Plate 1 (a) and (c) Gastropod mollusc shells of contrasting shape.
(b) Fossil shell of an ammonite.

Plate 2 Pacific bonito (*Sarda chiliensis*). (From MAGNUSON and
PRESCOTT (1966))

was in equilibrium with the solution. Thus the partial pressure of dissolved oxygen in water which is in equilibrium with the atmosphere is 0.2 atmosphere.

A gas tends to travel by diffusion from regions where its partial pressure is high to regions where it is low. In this way inequalities of partial pressure tend to be evened out. The rate of diffusion, $\mathcal{J}$ cm³/hr, depends on the gradient of partial pressure, dp/ds atm/cm (this is a mathematical expression meaning the gradient of a graph of partial pressure p against distance s). Let the area of the surface across which the diffusion is occurring be A cm². Then

$$\mathcal{J} = -AK\, dp/ds \qquad (2.1)$$

The negative sign indicates that diffusion occurs down the gradient. K is a quantity known as the permeability constant. Its value for oxygen diffusing through water is 2×10^{-3} cm²/atm hr and for oxygen diffusing through frog muscle 8×10^{-4} cm²/atm hr (WEIS-FOGH, 1964).

2.3   Animals with no circulatory system

We are now ready to discuss diffusion of oxygen in animals. Consider an animal which has no circulatory system but relies for its oxygen supply on diffusion through its tissues. Let all parts of its body use oxygen at the same rate, m cm³ oxygen/cm³ tissue hr. Let it be a long cylindrical worm-like animal, of radius r cm. A portion of its body, of length l cm, is shown in Fig. 2–3. Consider diffusion across the cylindrical surface of radius x cm, within the animal, which is indicated in the Figure. This surface has area $A = 2\pi x l$ cm². It encloses a volume $\pi x^2 l$ cm³ so the rate at which oxygen must be diffusing across it, to supply the tissues inside, is

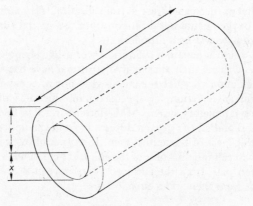

Fig. 2–3 This diagram is explained in the text.

$\mathcal{J} = \pi x^2 lm$ cm^3/hr. Putting these values and $s = (r-x)$ in equation 2.1,

$$\pi x^2 lm = -2\pi x l K \, dp/d(r-x)$$

$$dp/dx = mx/2K \tag{2.2}$$

Let the partial pressure of oxygen in the air or water in which the animal is living be p_e atm, and let the partial pressure of oxygen at the central axis of the animal be p_O. By means of the mathematical process of integration we can find out what difference of partial pressure $(p_e - p_O)$ is necessary if the partial pressure gradient across each layer of the body is to have the value given in equation 2.2. We find

$$p_e - p_O = \int_0^r mx \, dx/2K$$

$$= mr^2/4K \tag{2.3}$$

and since p_O cannot be negative

$$p_e \geqslant mr^2/4K$$

$$r \leqslant 2\sqrt{Kp_e/m} \tag{2.4}$$

The permeability constant K for oxygen diffusing in any tissue is likely to be fairly close to the value for frog muscle, 8×10^{-4} cm^2/atm hr. The partial pressure of oxygen, p_e, in the atmosphere or in well-aerated water is 0.2 atm. Putting these values in equation 2.4 we find that the greatest possible radius for a cylindrical animal relying on diffusion through its tissues for its supply of oxygen, is $0.026/\sqrt{m}$ cm. This maximum radius is plotted against the rate of oxygen consumption, m, in Fig. 2–4. It appears that if the animal uses about 0.1 cm^3 oxygen/cm^3 tissue hr (as do many animals, including the flatworms we will be discussing shortly, Fig. 2–2) its radius cannot exceed about 0.08 cm. This conclusion depends on the assumption that oxygen can diffuse in from all sides. If the animal rested on solid ground oxygen could not diffuse in from the ventral surface and the maximum radius would be a little less.

Only some animals have the cylindrical worm-like shape we have considered so far. Others, including many flatworms, have flattened leaf-like bodies. A calculation very like the one used to obtain equation 2.4 can be used to obtain an equation for the maximum thickness of a flattened animal with no circulatory system. If the lower surface of the animal rests on solid ground and all the oxygen diffuses in from the upper surface, we will call the thickness of the animal s cm. If oxygen diffuses in from both surfaces, we will call the thickness of the animal $2s$ cm. Thus s is the length of the diffusion path from the surfaces where oxygen is available to the tissue furthest from them. The equation is

$$s \leqslant \sqrt{2Kp_e/m} \tag{2.5}$$

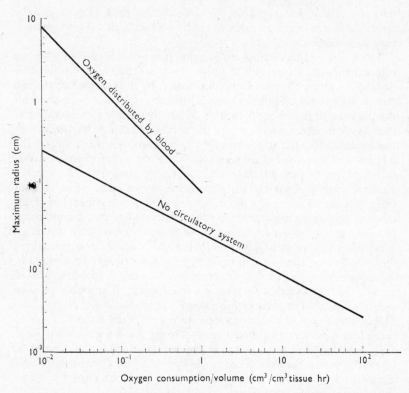

Fig. 2-4 Graphs of maximum radius against oxygen consumption for cylindrical animals with no circulatory system (calculated from equation **2.4**) and for ones with blood which distributes oxygen but without specialized respiratory organs (equation **2.6**, $d = 0.003$ cm).

If we give K and p_e the same values as before we find that the maximum value of s for an animal using 0.1 cm³ oxygen/cm³ tissue hr is 0.06 cm.

We will now look at the dimensions of some animals without circulatory systems. Protozoa are too small for diffusion of oxygen to pose any problem, at least when they live in well-aerated water, but the flatworms (phylum Platyhelminthes) are interesting. They apparently depend entirely for their oxygen upon diffusion through their tissues. Most of the free-living species live in streams and lakes and on shores in water which is well aerated. They creep over stones etc., so oxygen presumably diffuses in only from the dorsal surface. Most have flattened bodies. Our calculations indicate that they should not be more than about 0.06 cm thick, and this seems to be generally true. The smallest tend to be most nearly cylindrical and the largest to be flattest. Even a specimen 0.7 cm wide of a large shore-living flatworm, *Cycloporus papillosus*, is only 0.06 cm thick (Fig. 2–5). The larger flatworms could probably not exist without a circulatory system if they were thicker.

The flukes and tapeworms are parasitic flatworms. Some flukes live as external parasites on the gills of fishes. They depend on oxygen for their metabolism and are kept well supplied with it by the water which the fish pumps over its gills. Their thickness is limited by the same factors as limit the thickness of free-living flatworms. Most other flukes are internal parasites living in environments where the partial pressure of oxygen is low. For instance, the adult stage of the liver fluke, *Fasciola hepatica*, lives in the bile ducts of sheep and other mammals where the partial pressure of dissolved oxygen is probably generally about 0.03 atmosphere. A medium-sized specimen is about 0.08 cm thick and as oxygen could diffuse in from either surface of the body the maximum diffusion distance s is about 0.04 cm. By putting these values in equation **2.5** we find that the maximum possible rate of oxygen consumption, m, is only 0.03 cm³/cm³ tissue hr. This seems too low to supply by aerobic metabolism the energy needs of the animal since liver flukes in well-oxygenated water have been found to use oxygen very much faster. In experiments at low partial pressures of oxygen, and presumably in their natural habitat, liver flukes practise anaerobic metabolism, converting glycogen to propionic and acetic acids. Flukes which can do without oxygen are naturally not limited in size or shape by any necessity for the diffusion of oxygen, and a few grow to large sizes. *Hirudinella* is a fluke which lives in the stomachs of large fish, where the partial pressure of oxygen is low. It is not flattened but club-shaped, and grows to at least 1.5 cm diameter. Its metabolism is presumably almost entirely anaerobic.

Digestion in flatworms occurs largely in vacuoles in the cells of the gut wall. It seems likely that the resulting sugars, amino acids, etc. travel from there to the tissues that need them by diffusion, though it is conceivable that they might be passed from cell to cell by some process of active secretion. Certainly flatworms seem designed to allow diffusion of foodstuffs at adequate rates. The gut is relatively simple in the small cylindrical species but greatly branched in the large flattened ones, so that every part of the body is near a branch of the gut. The diffusion distances for foodstuffs are thus kept comparable to the diffusion distances for oxygen.

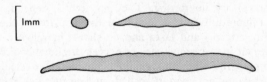

Fig. 2–5 Transverse sections of a few free-living flatworms, all to the same scale.

Tapeworms live in the guts of vertebrates but have no guts themselves. They absorb food by an active process but, once absorbed, it is very probably distributed to their tissues by diffusion alone. This may be why they are flattened like tapes. Their flatness is probably not necessary for the supply of oxygen since their metabolism seems to be mainly anaerobic.

Many sea anemones and jellyfish might seem to be exceptions to the rules we have been formulating. They have no blood circulation, their metabolism is aerobic and yet they are often large. How can this be explained? The outer surfaces of a jellyfish are covered by thin layers of cells but the interior of the animal is mesogloea, a jelly which contains few cells and so needs very little oxygen. A jellyfish such as *Cyanea* may be

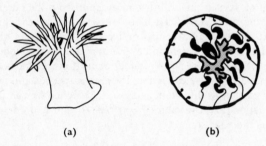

(a)   (b)

Fig. 2–6 Side view (a) and transverse section (b) of a sea anemone. The cavities which appear separate in this section interconnect nearer the base of the anemone.

50 cm in diameter and 5 cm thick, and yet nearly all its cells may be very close indeed to the surface of the body (or to the cavities and canals of the gastrovascular system, through which water is circulated). Sea anemones also have mesogloea, but have much less of it. Their bodies have thin walls enclosing a central cavity filled with seawater. Vertical partitions extend radially into this cavity from the body wall (Fig. 2–6). Included in them are the main muscles which contract to pull down the crown of the anemone when it closes up to protect itself from danger. The mouth connects the central cavity to the sea, and the water in the cavity is changed continuously by the inward currents driven by cilia at one or both ends of the mouth. This water can circulate between the partitions. The partitions are thin even in large anemones. For instance, they are 0.1 cm thick or less in a specimen 8 cm in diameter of *Tealia felina*. The outer wall of the body is thicker but much of its thickness is occupied by mesogloea. Nearly all the cells are within 0.05 cm either of the external water or of the water in the central cavity, so provided the water in the central cavity is changed fast enough to prevent the partial pressure of its dissolved oxygen from falling too low, diffusion should be able to provide the tissues with plenty of oxygen.

2.4 Animals with blood which distributes oxygen

Now we will consider animals which circulate blood round their bodies, and use this blood to collect oxygen at the surface of the body and deliver it to the tissues. Most of them have a respiratory pigment such as haemoglobin, which increases the quantity of oxygen which can be carried by a given volume of blood. We will consider a worm-like animal with no specialized respiratory organs which takes up oxygen through the whole of its outer surface. Is there any limit to the size of such an animal? Let it be cylindrical with length l cm and radius r' cm. Its volume is then $\pi r'^2 l$ cm^3 and if the rate of consumption of oxygen is m cm^3/cm^3 tissue hr, the amount of oxygen used in an hour by the whole animal will be $\pi r'^2 lm$ cm^3. This amount of oxygen must diffuse in an hour from the outer surface of the body to the most superficial blood vessels, a distance which we will call d cm. If the partial pressure of oxygen is p_e atm in the external medium (whether air or water) and averages p_b atm in the blood, there is a gradient of partial pressure $(p_b - p_e)/d$ atm/cm. If our cylindrical animal is long and relatively thin we can ignore the area of its ends and take the area available for oxygen to diffuse in as $2\pi r' l$ cm^2. Hence we can put into the diffusion equation (2.1) the values $\mathcal{J} = \pi r'^2 lm$, $A = 2\pi r' l$, $\mathrm{d}p/\mathrm{d}s = (p_b - p_e)/d$ This gives us the equation

$$r' = 2K(p_e - p_b)/md \qquad (2.6)$$

We will take $K = 8 \times 10^{-4}$ and $p_e = 0.2$ as before. p_b seems unlikely to be much less than about 0.05: it would have this value if, for instance, the partial pressure of oxygen was 0 in the blood arriving at the surface and 0.1 atm in the blood leaving the surface after taking up oxygen. We will therefore assume that $(p_e - p_b) \leqslant 0.15$. We still need a value for d.

The animal we are considering, which is cylindrical and has circulating blood but no specialized respiratory organs, might well be an earthworm. The cuticle and epidermis of a typical earthworm are together about 0.005 cm thick. The most superficial blood vessels form loops within the epidermis (LAVERACK, 1963) and 0.003 cm seems a reasonable estimate of the diffusion distance d. If the blood were brought much closer to the surface than this, the animal would probably be rather easily damaged. Blood is brought much nearer to the surface in the gills of fishes, but these are delicate structures in a protected position.

Putting the values we have selected in equation 2.5 we find $r' \leqslant 0.08/m$. This maximum radius is plotted against m in Fig. 2–4. It is clear that for a given rate of consumption of oxygen m, a much larger radius is possible for an animal in which oxygen is transported by circulating blood than for one which depends entirely on diffusion. Earthworms can and do grow much larger than the cylindrical types of flatworm. The largest tropical earthworms which have been investigated have been found to use about 0.06 cm^3 oxygen/g hr (Fig. 2–2) and Fig. 2–4 indicates that the maximum radius for this oxygen consumption is 1.3 cm. This is about the radius of the largest earthworms, such as *Rhinodrilus fafner* of South America. However, the diffusion distance between atmosphere and blood may not be the same as the distance of 0.003 cm, derived from smaller worms, on which the graph is based. Also the giant worms are not only fat but are long and correspondingly heavy. The largest must weigh almost 1 kg. They probably use oxygen at less than the rate of 0.06 cm^3/g hr, which is for worms weighing only 18 g.

The limit we have just estimated does not of course apply to animals which use gills or lungs. The gills of many animals (including fish) occupy protected positions, and all lungs are protected by being inside the animal, so blood can safely be brought very close indeed to the water around the gills or the air filling the lungs. Also, the areas of the surfaces through which oxygen diffuses can be made extremely large. The gills of fish have so complicated a structure that their surface area is generally about double the area of the external surface of the body. Lungs have a complicated pouched structure and it has been estimated that the area available for diffusion in a man's lungs is at least 50 m^2. Animals which have gills or lungs can grow very large, and factors other than the availability of oxygen are responsible for any limitation to their size.

2.5 Animals with tracheae

Insects have a different system for getting oxygen to the tissues which need it. In most of them the blood has no respiratory pigment and plays no part in respiration. Its main functions are the transport of foodstuffs, waste products, and hormones, and as a hydrostatic fluid that transmits the pressures needed for certain movements. Oxygen travels to the tissues through tracheae. These are air-filled tubes which start at holes called spiracles at the surface of the insect and penetrate the tissues, dividing as they go into finer and finer branches (Fig. 2–7(a)). Oxygen diffuses along them, from the atmosphere to the tissues which need it. This diffusion of gaseous oxygen through the tracheae is very much faster than diffusion of dissolved oxygen through the tissues could be. The permeability constant for oxygen diffusing through air is 660 cm^2/atm hr, or 800 000 times the value for diffusion through tissue. Of course the tracheae only comprise a small fraction of the body volume. A section of insect tissue cut at right-angles to the tracheae will often have only 1 per cent of its area occupied by tracheae, but even this allows diffusion to proceed at 8000 times (i.e. 1 per cent of 800 000 times) the rate which would be possible if there were no tracheae (WEIS-FOGH, 1964). This assumes that the partial pressure of oxygen in the tissues is the same as in the nearest trachea.

Let us consider how this affects the maximum size of the animal. Consider first a caterpillar. This is a more or less cylindrical animal and we might consider modifying equation 2.4 to get an estimate of maximum

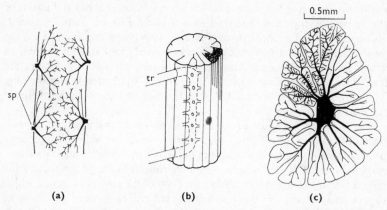

0.5mm

(a)   (b)   (c)

Fig. 2–7 (a) Dorsal view of part of a dissection of a caterpillar, showing tracheae branching over the gut. sp, Positions of spiracles. (b) A diagram showing the path of the main trachea (tr) through a common type of insect flight muscle. (c) Transverse section of an insect flight muscle of the type illustrated in (b), showing smaller tracheae radiating out from the main central one. (c redrawn from WEIS-FOGH, 1964)

radius. This cannot be done at all realistically, however, since the tracheae do not open all over the surface of the animal but only along its sides (Fig. 2–7(a)). We will take a different approach, using data on goat-moth caterpillars (WEIS-FOGH, 1964, and papers cited by him). A goat-moth caterpillar weighing 3.4 g was found to have tracheae of mean length 0.74 cm (measured from their inner ends to the nearest spiracle), so if the difference of partial pressure of oxygen between the atmosphere and the tissues was (p_e-p_0) atm the gradient of partial pressure would be $-(p_e-p_0)/0.74$ atm/cm. The total cross-sectional area of all the tracheae, measured close to the spiracles, was 0.06 cm^2 but the tracheae of caterpillars taper so an average total value for the whole length of the tracheae might be 0.03 cm^2. The oxygen consumption of a goat-moth caterpillar of this size, crawling slowly around, is about 1 cm^3/hr (0.3 cm^3/g hr). Putting these values, and the permeability constant for oxygen in air, in equation **2.1** we find

$$1 = 0.03 \times 660\,(p_e-p_0)/0.74$$

$$p_e-p_0 = 0.04 \text{ atm}$$

(Calculations concerning diffusion through the stomata of plants have to take account of diffusion in the air outside the stomata as well as in the pores themselves (SUTCLIFFE, 1968), but this is not necessary in calculations on tracheae because they are so long that most of the change in partial pressure between the atmosphere and the tissues occurs within them.)

Goat-moth caterpillars burrow in the wood of trees, and their burrows are probably poorly ventilated, but it would obviously be possible for a caterpillar living in air containing 0.2 atm oxygen to grow much fatter and still obtain an adequate supply of oxygen through its tracheae. A 3.4 g goat-moth caterpillar would have a radius of about 0.4 cm, and it can be estimated that a radius of 0.9 cm would be possible for a caterpillar with the same oxygen consumption (0.3 cm^3/g hr) and the same proportion of its volume occupied by tracheae. Figure 2–4 indicates that the maximum radius for an earthworm-like animal with the same oxygen consumption is only 0.3 cm.

We will now consider adult insects. Not only do they use oxygen very fast when they fly (Fig. 2–2), but most of the oxygen is used by the flight muscles which constitute only a small proportion of the body volume. These use oxygen at enormous rates which vary between species but are usually around 100 cm^3/cm^3 tissue hr. It would be quite unrealistic to assume, as has been done in previous calculations, that all the tissues in the animal use oxygen at the same rate.

Flight muscles have their tracheae arranged in various ways, and a typical arrangement is shown in Fig. 2–7(b) and (c). A large trachea enters the muscle near one end, runs along the length of the muscle and leaves at the other end. Branches radiate out from it all along its length and divide into smaller branches which permeate the whole muscle. The finest branches form a network investing and sometimes penetrating into the individual muscle fibres. Small insects such as the fruit fly *Drosophila* seem to depend entirely on diffusion for the transport of oxygen from the spiracles to the muscle fibres, but diffusion would not be fast enough along the greater distance in larger insects such as bees and dragonflies. These pump air through their main tracheae including the central tracheae of flight muscles, but transport along the smaller tracheae including the branch tracheae in the muscles is still by diffusion alone. The muscles therefore cannot be very thick if they are to be capable of using oxygen at high rates. Limits can be calculated by considering the rate of diffusion of oxygen in essentially the same way as we have estimated limits for the sizes of animals with various respiratory arrangements. We will assume a very large central trachea, of diameter 0.2 of the diameter of the muscle. We will assume that the smaller tracheae are very large and plentiful, occupying 10 per cent of the area of any section cut at right-angles to them. It can be calculated that the greatest diameter such a muscle could have and still be able to consume 100 cm^3 oxygen/cm^3 tissue hr is about 0.5 cm. The only insect flight muscles known to exceed this diameter have a peculiar arrangement of tracheae, so that air is pumped through the main branch tracheae as well as the central one. They are found in *Lethocerus*, a genus of tropical water bugs which are among the largest insects, reaching 11 cm in length (WEIS-FOGH, 1964; see Plate 4).

The muscles of insects are not permeated by blood capillaries like those of vertebrates, but are bathed in the blood which fills the body cavity. While oxygen diffuses into the muscle from the central trachea, food-stuffs diffuse in from the blood outside. The muscles are not compact cylinders but are divided, as Fig. 2–7(c) shows, into small groups of fibres. The blood in the spaces between these groups of fibres is continually being changed, since it is alternately squeezed out and drawn in as the muscle contracts and relaxes. Consequently the greatest distance food-stuffs have to diffuse from the blood to any muscle fibre is far less than the radius of the muscle.

2.6   Heat loss by birds and mammals

We have seen in previous sections how diffusion of oxygen may set upper limits to the radii or thicknesses of certain types of animals. We will see in this one that loss of heat may set a lower limit to the size ranges of birds and mammals.

Birds maintain their bodies at constant temperatures of about 40°C and mammals maintain theirs at about 37°C. Since chemical reactions proceed faster at high temperatures a constant high temperature enables an animal with appropriately adapted enzymes to be highly active at any time, how-ever the temperature of the environment may change. If the body is kept warmer than the environment heat must be lost to the environment. This can be reduced by insulation. The feathers of birds and the fur of mammals are insulating coverings. Air is trapped between the feathers or hairs so that it cannot easily circulate: convection is inhibited, and most of the heat lost through this air travels by conduction. Trapped air has good heat-insulating properties because the thermal conductivity of air is much lower than those of most solids and liquids.

All metabolic processes release heat, and the heat released by processes required for purposes other than heat production may suffice to maintain the temperature of the body. Alternatively, additional metabolism may be necessary for the sole purpose of producing the required heat. This is particularly likely for small animals because not only is the ratio of the area from which heat is lost to the volume of the body engaged in producing heat greater for them than for larger animals, but they cannot conveniently carry as great a thickness of insulation. It seems likely that there is a mini-mum size below which so much additional metabolism (and so food) would be necessary that a warm-blooded animal would be at a disadvan-tage. This may explain why there are no birds smaller than the smallest hummingbirds (about 2 g) or mammals smaller than the smallest shrews (also about 2 g). We will discuss this quantitatively, referring only to mammals although the same principles apply to birds.

We will ignore loss of heat by evaporation which may be large when a mammal is sweating or panting but is relatively small when it is conserving heat. The rest of the heat loss occurs in two stages, by conduction through the fur followed by radiation and convection from the outer surface of the fur. We will estimate the amount of heat that is lost by considering conduction through the fur.

Conduction occurs down temperature gradients just as diffusion occurs down gradients of partial pressure, and is described by an equation very like the diffusion equation **2.1**. If heat is being conducted at a rate H J/hr over an area A cm^2 down a temperature gradient dT/ds °C/cm

$$H = -AC \, dT/ds \qquad (2.7)$$

where C is the thermal conductivity of the material the heat is travelling through, in J/cm hr °C.

We will consider a mammal which we will suppose to be spherical. This will tend to make us underestimate heat loss since a spherical surface is the smallest which can enclose a given volume, but it is not as unrealistic as might be thought since limbs are generally kept cooler than the rest of the body and so do not lose as much heat as their surface area suggests. Let the animal have radius r cm so that its volume is $4\pi r^3/3$ cm^3. Let it consume m cm^3 oxygen/cm^3 tissue hr. Then, since metabolism involving 1 cm^3 oxygen releases about 20 J whatever food is being oxidized the rate H at which heat is being produced and lost by the whole body is 80 $m\pi r^3/3$ J/hr. Let the thickness of the fur be 0.3r cm: this value is based on measurements on shrews and on small and medium-sized arctic animals (SCHOLANDER, 1955), and is about the largest value we could reasonably assume. Longer fur would be apt to hinder walking. Hence the area of the skin is $4\pi r^2$ cm^2 and the area of the outer surface of the fur is $4\pi(1.3r)^2$ cm^2. We will take an intermediate value, $5\pi r^2$ cm^2, as the area A. The conductivity of fur is about 1.6 J/cm hr °C (SCHOLANDER, 1955. This is rather less than twice the conductivity of air). If the temperature of the body is T_b °C and that of the outer surface of the fur T_s °C the temperature gradient is $-(T_b - T_s)/0.3r$ °C/cm. Putting these values in equation **2.7** we find

$$80m\pi r^3/3 = 5\pi r^2 1.6(T_b - T_s)/0.3r$$

$$m = (T_b - T_s)/r^2$$

Since the density of mammals is about 1 g/cm^3 the weight W of our mammal is about $4\pi r^3/3$ g, $r^2 \simeq (3W/4\pi)^{0.67} = 0.4W^{0.67}$ and

$$m \simeq 2.5(T_b - T_s)W^{-0.67} \qquad (2.8)$$

This equation gives an estimate of the oxygen consumption of a very well insulated mammal which is not sweating, but is maintaining its body at T_b °C in conditions which keep the outer surface of the fur at T_s °C. It gives a minimum value for the oxygen consumption per unit weight of a mammal of given weight. Notice that it is proportional to $W^{-0.67}$.

Figure 2–8 shows actual rates of oxygen consumption per unit weight,

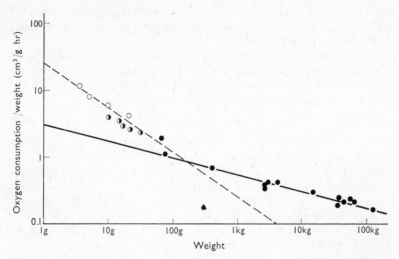

Fig. 2–8 Rate of oxygen consumption per unit body weight plotted against body weight for: ○, various species of shrew at 24°C; ◑, species of mouse at 24°C; and ●, other mammals up to the size of a pig, at temperatures mostly between 24 and 28°C. The continuous line represents proportionality of oxygen consumption/weight to $(weight)^{-0.25}$. The broken line indicates proportionality to $(weight)^{-0.67}$, and has been calculated from equation **2.8** for $(T_b - T_s) = 10°C$. (Data from PEARSON (1947, 1948) and BENEDICT (1938))

for resting mammals of various sizes. The rates for mammals weighing 50 g or more are roughly proportional to $W^{-0.25}$ as for animals in many other groups (Fig. 2–2). The points for mice and shrews, however, lie in lines much more nearly parallel to the broken line of gradient −0.67 which has been calculated from equation **2.8**. The rates for mice and shrews were determined in conditions in which $(T_b - T_s)$ probably lay between 5 and 10°C and the line assumes a value of 10°C, so agreement with equation **2.8** is reasonably good.

It thus appears that the metabolism of medium and large mammals releases a good deal more heat than the amount (indicated by the broken line) which would be necessary to maintain their body temperature if their fur was as thick in proportion to body diameter as the fur of small mammals. Maintenance of the body temperatures of mice and shrews, however, requires faster metabolism than would be predicted by extrapolation along the continuous line from the oxygen consumptions of larger mammals. Mammals which were smaller still might have to obtain impossible quantities of food to support the very fast metabolism which they would need. The smallest shrews and hummingbirds are probably about as small as it is feasible for mammals and birds to be, in environments where the temperature falls more than a few degrees below their body temperatures.

3.1 Shapes for catching light and food

Most plants produce the main materials they need for metabolism and growth by photosynthesis, and natural selection must generally favour characters which increase the amount of light which falls on the plant. No more light falls on the foliage of a forest than would fall on an equal area of grassland in the same conditions of weather and season, but tall plants have evolved because height gives an individual plant an advantage over the shorter plants which it overshadows. Similarly the outer leaves of a tree may shade inner ones so that little light is available to them for photosynthesis. If the foliage were thick throughout the crown of a tree the leaves at the centre might even receive so little light that they destroyed more food by metabolism than they produced by photosynthesis. In fact trees often have the form shown in Fig. 3–1, with an outer shell of foliage enclosing a core of relatively leafless branches.

Fig. 3–1 Diagrammatic section of a tree showing an outer shell of foliage (stippled) enclosing a relatively leafless core.

Most plants have evolved thin leaves because thick leaves require more material for their construction and use more energy in their metabolism than thin ones of the same area, but receive no more light. They are often arranged in a neat mosaic, so as to form a more or less continuous curtain of foliage with remarkably little overlap of individual leaves. Little light is then lost by passing between the leaves, and at the same time there is little overshadowing of one leaf by another of the same plant (Fig. 3–2).

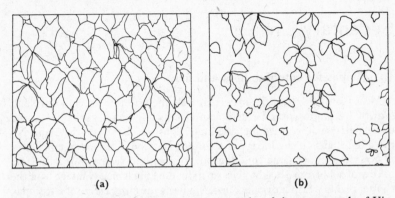

(a)   (b)

Fig. 3–2 Outlines traced from two photographs of the same patch of Virginia creeper (*Parthenocissus* sp.), growing on a wall. (a) shows the intact patch. All the leaves and parts of leaves which are visible were exposed to direct sunlight. (b) shows leaves and parts of leaves which were shaded. They were identified by spraying the creeper with paint, which covered the exposed leaves but left the shaded ones green. The exposed leaves were removed so that the shaded ones could be photographed.

Many animals do not move about, but occupy a fixed position like plants. Such sedentary animals are shaped for catching food, not light, but their shapes are often rather plant-like. They include the stalked crinoids or sea-lilies, which are relatives of the starfishes, and live anchored by their stalks to the bottom of the sea, mostly at depths between 200 and 5000 m (Fig. 3–3(a)). There are also free-swimming, stalkless crinoids which live in shallower water. Crinoids have a central body with a mouth in the middle of a disc formed by branched arms. The arms and their side-branches (pinnules) bear sticky projections known as tube feet which are too small to be shown in the Figure. The shape of stalked crinoids suits their manner of feeding. They eat such particles of (mainly dead) organic matter as sink onto their arms from the plankton living nearer the surface. The tube feet are just long enough to prevent particles from falling between one pinnule and the next. The particles are caught by the tube feet, passed by them to grooves on the pinnules and arms and passed along the grooves to the mouth by the action of cilia in the grooves. A disc of arms seems the ideal arrangement. If the arms were longer but did not form a complete disc, they might cover the same area and catch the same amount of food, but the average distance which the food would have to travel along the grooves to the mouth would be greater and the food would be more likely to be lost, or stolen by the crinoid's parasites. Also, the longer arms would have to be thicker at their bases, to be sufficiently strong and stiff, so more material would be needed to construct them.

Most stalked crinoids probably live in slow-moving or still water.

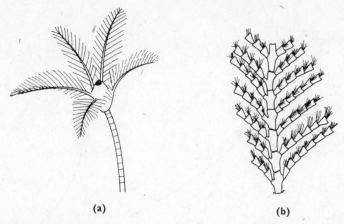

(a)   (b)

Fig. 3-3 Sketches of (a) a typical crinoid and (b) the hydroid *Aglaophenia*.

Many other sedentary marine animals live closer inshore where they are exposed to faster water movements due to waves and tides. Among them are *Aglaophenia* (Fig. 3-3(b)) and other colonial hydroids. Rather than arms or branches spreading out in a horizontal plane to receive food falling from above they need branches spread out in such a way as to catch as much as possible of the food carried by the current. Many hydroids which live in places where the direction of flow is variable have three-dimensional bush-like colonies, but in places where the flow due to waves or tides is backwards and forwards in a single direction, species with flat colonies like that of *Aglaophenia* predominate (RIEDL, 1964). They grow with the plane of the colony at right angles to the prevailing current. The spacing of the branches and of the individual hydroids on the branches is such that when the individuals have their tentacles extended they can just prevent food from passing between them. This is as neat an arrangement for catching food as the mosaic of plant leaves is for catching light.

The shape of such colonies must make the water tend to twist them into the ideal position, at right-angles to the current. Consider such a colony, or any other symmetrical flexible blade with a central midrib. If it is held obliquely in a current the parts on either side of the midrib will bend as shown in Fig. 3-4(a). Side A bends so as to be more nearly at right-angles to the current and side B so as to be more nearly parallel to it. The water thus exerts a greater force on A than B and tends to twist the colony until it stands symmetrically across the current (Fig. 3-4(b)). This is probably how the group of soft corals known, from their shape, as sea fans (*Gorgonia* spp.) get twisted. They start their life arranged at random in all directions and are gradually twisted as they grow until they stand at right-angles to the current (WAINWRIGHT and DILLON, 1969).

Crinoids catch food which sinks onto them and colonial hydroids have

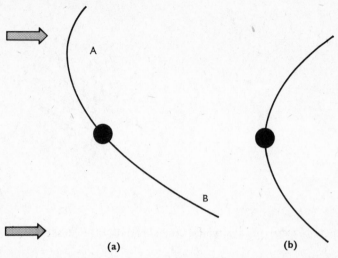

Fig. 3-4 Diagrammatic horizontal sections of a flexible blade with a central midrib, set (a) obliquely and (b) transversely in a current. Further explanation appears in the text.

their food brought to them by natural water movements. Sponges pump water through passages within themselves, and filter out the food particles it contains. The velocity of the water flowing through the fine filters is necessarily low, but if the sponge lives in still water it is desirable that the filtered water should be ejected at high velocity so that it travels some distance and is not immediately drawn in for re-filtration. This has been achieved by the evolution of sponges with many entrances and one or a few small exits. The entering water travels slowly through the numerous passages within the sponge which are lined by the cells which drive and filter it. The passages join and the water has to leave the sponge through an aperture whose cross-sectional area is far less than their aggregate cross-sectional area so it leaves as a relatively fast jet. For instance, water enters the sponge *Leuconia* at about 0.1 cm/s and slows down considerably as it passes through the filters, but leaves at about 8.5 cm/s (NICOL, 1967).

3.2 Sizes and shapes of leaves

The arrangement of leaves on trees has been discussed but not their shapes and generally quite small sizes. Some plants have enormous leaves but the great majority even of large trees have leaves measuring less than about 0.1 m across. Larger leaves such as those of sycamore are often deeply lobed. These observations may perhaps be explained by the danger of overheating in sunlight.

A leaf in full sunlight absorbs far more radiant energy than it could lose by re-radiation unless its temperature rose far above the lethal limit. It can

Plate 3 A blowfly and an Alsatian dog, showing the difference in posture. (Photographs by Mr. A. O. Holliday)

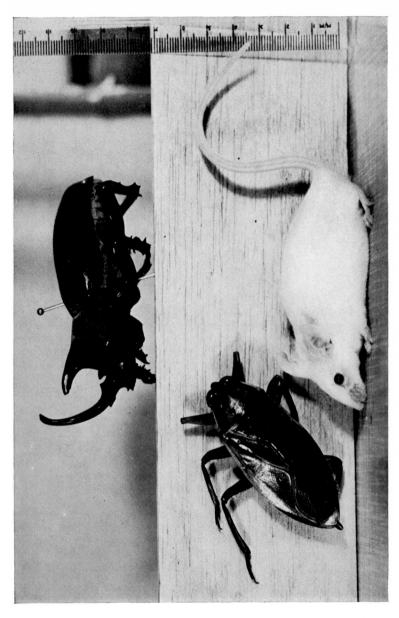

Plate 4 Two of the largest insects, with a mouse for comparison of size. *Above* a rhinoceros beetle and (*left*) a giant water bug. An even larger water bug, 11 cm long, is referred to in Chapter 2.

survive because it is cooled by convection and by the evaporation of the water it transpires. Even so it may get very hot: GATES (1965) found that sunlit oak leaves were at 49°C on an occasion when the temperature of the air around them was only 28°C. When water is short, convection is particularly important, and convection may be slow if wind is slight or absent.

When air driven by wind or by local convection currents flows over a leaf, the air immediately in contact with the leaf remains stationary and the air near its surface is slowed down. This slow-moving air forms the boundary layer which is referred to again in Chapter 4. It is very thin near the windward edge of the leaf but gets progressively thicker with distance from that edge. Thus heat leaving the downwind part of a large leaf has to pass by conduction through a relatively thick layer of almost still air before being carried away by the moving air. Also, the air passing this part of the leaf has already been warmed by the rest of the leaf. Consequently the rate of loss of heat from the downwind parts of large leaves tends to be rather low. A given area of foliage will be able to lose heat faster if it is divided into many small leaves rather than a few large ones. The same advantage can be obtained by dividing large leaves into deep lobes so that air travelling over them in most directions does not encounter a long unbroken surface. These considerations are only important to leaves liable to be exposed simultaneously to full sunlight and almost still air.

Tests on copper sheets of various shapes, cooled by slow currents of air, have confirmed that deeply lobed sheets do in fact lose heat a little faster than discs of equal area (VOGEL, 1970). It has even been possible to show that the small, deeply lobed leaves which grow on the parts of oak trees which are exposed to strong sunlight lose heat faster (per unit area) than the larger less-deeply lobed leaves which grow in shady positions (Fig. 3–5).

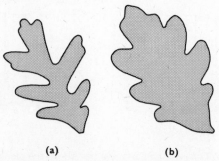

(a)   (b)

Fig. 3–5 Outlines of leaves of white oak (*Quercus alba*) used in experiments on heat loss by convection. (**a**) is a typical leaf from a part of the tree exposed to strong sunlight and (**b**) is a typical leaf from a shaded part. (From VOGEL, 1970)

If a leaf were to become progressively more and more deeply lobed the ₁obes would eventually become completely separated so that the leaf became a compound one like the leaves of ash, consisting of several small leaflets. It would be tempting to suppose that compound leaves evolved in this way, due to selection for rate of loss of heat, were it not that other evidence points in the opposite direction, indicating that large simple leaves evolved from compound ones (CORNER, 1964).

3.3 The thickness of tree trunks

A tree trunk must be strong enough to withstand the forces which act on it. The principal forces are the weight of the tree (which acts vertically) and the drag exerted on it by the wind (which acts horizontally). Neither is easy to measure for any but the smallest trees, but both have been measured for conifers of various species, 8 m high, in experiments at the Royal Aircraft Establishment, Farnborough (FRASER, 1962). The trees were fixed in a large wind tunnel, and the drag was measured at various wind speeds. It was found to be equal to the weight of the tree (excluding roots) at about 40 mph (18 m/s). Trees in exposed situations have to endure 40 mph winds in gales, and may sometimes suffer gusts of much higher speed. They are thus likely to have to withstand drag forces rather greater than

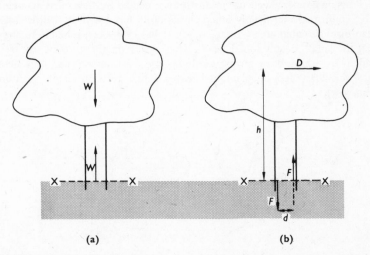

(a) (b)

Fig. 3–6 Forces acting on the parts of a tree above the plane XX. (a) Forces acting in still air. The wood of the trunk is compressed by the weight of the tree so that the part of the trunk below XX exerts an upward force W, equal to the weight, on the part above. (b) Additional forces acting in wind. The trunk is stretched on the windward side and compressed on the lee side so that the parts below XX exert on the parts above the forces F which balance the moment due to the drag D.

their own weight. However, a drag equal to the weight sets up very much greater stresses in the trunk than the weight does, as we shall see.

Consider a tree (Fig. 3–6(a)) whose weight, excluding parts below XX, is W. If the centre of gravity is vertically above the trunk there will be no tendency for the weight to bend the trunk, and the only force transmitted through the trunk at XX will be the vertical force W. Now suppose the wind exerts on the tree a drag D which acts at a height h above XX (Fig. 3–6(b)). This drag tends to bend the trunk and it exerts a moment Dh (referred to as the bending moment) which stretches the wood of one side of the trunk at XX and compresses the wood of the other side. The forces of tension and compression at XX must balance the bending moment, and if they act at a distance d apart they must have a value F so that

$$Fd = Dh \tag{3.1}$$

Since d cannot exceed the diameter of the trunk $h \gg d$ and $F \gg D$. The forces set up in the trunk by the drag are much greater than the drag itself. Drag equal to the weight of the tree would set up stresses in the trunk far greater than those due to the weight. Wind is much more important than weight in determining what thickness of trunk is necessary. The argument has been simplified by ignoring the possibility of elastic instability (ALEXANDER, 1968) but the conclusion is sound. It does not apply to horizontal branches since weight as well as drag acts at right-angles to them.

If the trunk is circular in cross-section and is made of wood of tensile strength T, the minimum radius r_{min} which it must have at XX if it is to withstand the bending moment Dh due to the drag is given by the equation

$$r_{min} = (4Dh/\pi T)^{\frac{1}{3}} \tag{3.2}$$

(ALEXANDER, 1968). The corresponding cross-sectional area A_{min} is πr_{min}^2 so

$$A_{min} = \pi^{\frac{1}{3}}(4Dh/T)^{\frac{2}{3}} \tag{3.3}$$

Notice that these equations imply that the trunk should taper, with its thickest part at the bottom where Dh is greatest. The equations will be used later but no attempt will be made to calculate from them the trunk thickness needed for a given size of tree. It would be hard to do this realistically because wind speeds vary from place to place and because trees bend so much in the wind that it is unrealistic to apply to them the standard equation for drag on rigid bodies (equation 4.1) (FRASER, 1962).

Since competition for light is often important in groups of trees, natural selection must tend to favour trees which grow high quickly. If the rate at which the materials for growth can be produced is limited, this implies trunks which are as slender as possible. On the other hand, trees which are easily broken by the wind must tend to be eliminated by natural selection. One might therefore expect trees to grow trunks thick enough for breakage to be unlikely, but not so thick that it can never happen. This is generally the case: trees are occasionally snapped by gales. Similar arguments apply to roots, and trees are occasionally uprooted by gales.

3.4   Angles of branching

Let us consider now how different patterns of branching affect the total amount of timber needed to form a tree of adequate strength. It seems reasonable to expect natural selection to favour trees whose pattern of branching reduces the timber required to a minimum, so that growth can be rapid.

Consider a side branch from a tree with a vertical main trunk (Fig. 3–7(a)). Suppose the tree must be strong enough to withstand a drag D which acts at a height h from the ground on the foliage of the main part of the tree, and a much smaller drag D' which acts on the foliage of the branch we are considering at a height h' and at a distance a to the side of the main trunk. Equation 3.3 can be used to obtain the requisite cross-sectional areas for the base of the branch and for the main trunk immediately above and below the point of branching. These can be used to find out how much the total volume of wood needed in the tree would be changed if the angle of branching θ were changed slightly without changing h, h' or a. This involves some rather tiresome calculus which will not be printed here. It

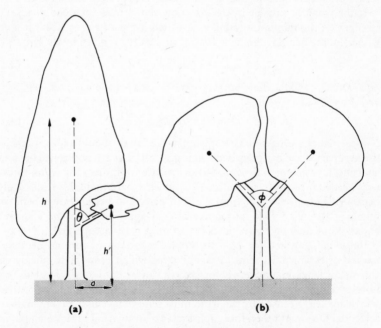

Fig. 3–7 Diagrams illustrating the discussion of branching angles.

leads to the conclusion that the volume of wood is least when $\theta = 90°$. It thus provides a possible explanation for the branching pattern of many conifers including larch (*Larix* spp.) and spruce (*Picea* spp.) which have their numerous side branches set more or less at right angles to the main trunk (Fig. 3–8(a)). However, many trees have relatively small side branches attached to the main trunk at angles θ less than 90°, and so must contain more wood than is necessary to carry their foliage. A particularly striking example is the Lombardy poplar (*Populus italica*), in which θ is generally about 20° (Fig. 3–8(b)).

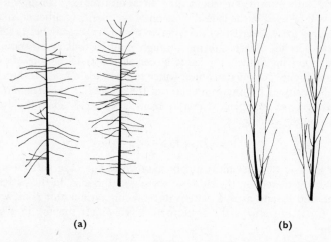

(a)   (b)

Fig. 3–8   Branching patterns of (a) larch and (b) Lombardy poplar, from photographs.

If similar arguments are applied to a tree whose trunk divides into two equal branches (Fig. 3–7(b)) one finds that least wood is needed if the angle of branching, ϕ, is 120°. However, trees with a divided main trunk tend to branch in so irregular a way that it seems obvious that they do not conform to any pattern which mathematics might suggest.

Swimming and Flight

4.1 Reynolds number

This chapter is about the shapes which enable animals to swim or fly with least expenditure of energy, about the ways in which the best shape depends on size and speed, and about the limits of size for flying animals. It starts with information about drag on bodies moving through fluids, which will be used in the later discussion of the merits of different shapes.

The drag forces exerted by the wind on trees have already been referred to. Drag acts on objects moving through fluid as well as on stationary objects in moving fluid and so has to be considered in discussions of swimming and flight. Consider an object which moves with velocity V through a fluid of density ρ. Let the frontal area of the object (i.e. the area of a full-scale drawing of the object as seen in front view) be A. Then the drag D is given by the equation

$$D = \tfrac{1}{2}\rho V^2 A C_D \tag{4.1}$$

where C_D is a dimensionless number known as the drag coefficient. C_D depends not merely on the shape of the object but also on the pattern of flow around it, and as this pattern depends on various things as well as shape C_D is not always the same for a given object moving in a given direction.

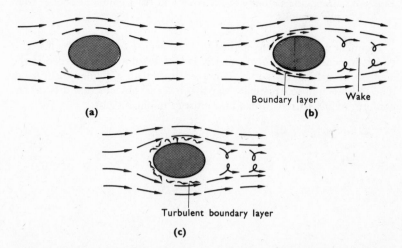

Fig. 4-1 Diagrams which are explained in the text, showing the flow of fluid around a stationary object at three different Reynolds numbers.

Three fundamentally different patterns of flow are possible, and are illustrated in Fig. 4–1. In (a) the fluid is flowing smoothly around the body, parting in front of it and closing up behind it. The fluid in contact with the body moves at the same velocity as the body and the velocity changes gradually with distance from the body. Thus the movement involves gradients of fluid velocity and is opposed by the viscosity of the fluid. Nearly all the drag is due to viscosity. In (b) and (c) the inertia of the fluid is important as a cause of drag, as well as viscosity. In (b) the fluid parts and flows smoothly over the front part of the body but fluid quite close to the body is moving at almost the same velocity as fluid far from it. The main gradient of velocity occurs in a thin layer of fluid, known as the boundary layer, next to the surface of the body. The fluid does not close up smoothly behind the body but forms a wake of swirling eddies. Part of the drag is due to the viscosity of the fluid in the boundary layer. The rest is due to the changed momentum of the fluid in the swirling wake; a force is required to change momentum, according to Newton's second Law of Motion. In (b) the flow in the boundary layer is smooth and parallel to the surface (laminar flow) but in (c) it is irregular (turbulent flow). There is a wake as in (b).

Which pattern of flow occurs around an object of given shape depends on the size of the object, on the velocity, and on the density and viscosity of the fluid. All these factors are incorporated in a dimensionless number known as the Reynolds number, which is very useful in the study of flow. If the fluid is water, the Reynolds number is

$$10^6 \text{ (length of object in m)} \times \text{(velocity in m/s)}$$

If the fluid is air, the Reynolds number is

$$7 \times 10^4 \text{ (length in m)} \times \text{(velocity in m/s)}$$

The drag coefficients of objects of the same shape are the same if the Reynolds number and direction of motion are the same. It does not matter if the fluid, the velocity and the size of the object are all different, nor whether it is the object or the fluid that is moving, provided only that shape, direction and Reynolds number are the same.

The pattern of flow shown in Fig. 4–1(a) occurs at Reynolds numbers below 1. As Reynolds number increases above 1 the pattern of flow changes gradually to (b), which is maintained until the number reaches a critical value somewhere between 2×10^5 and 2×10^6, at which there is a sudden transition to pattern (c). Pattern (c) persists at all Reynolds numbers above this. An animal must generally be very small and slow for flow of type (a) to occur around it, and large and fast for flow of type (c) to occur. An animal 1 mm long moving through water at 1 mm/s would have a Reynolds number of 1, while an animal 1 m long moving through water at 1 m/s would have a Reynolds number of 10^6.

4.2 Streamlining

In the range of Reynolds numbers in which they leave wakes (Fig. 4-1(b) and (c)) the drag on bodies of most shapes is mainly due to the changed momentum of the fluid in the wake. Drag can be greatly reduced if the body is shaped in such a way as to minimize eddying in the wake. The shape required is a streamlined one, rounded at the front and tapering to a point behind, like a torpedo without fins (Fig. 4-3(a)). Figure 4-2 shows the advantage of streamlining. The drag coefficient of a streamlined body is only 0.1 to 0.25 that of a sphere. The drag coefficient is calculated from the frontal area (equation **4.1**), but the volume of a streamlined body can be far more than that of a sphere of the same frontal area. Over a wide range of Reynolds numbers the drag on a streamlined body of the proportions shown in Fig. 4-3(a) is only about 0.05 to 0.1 that on a sphere of the same volume, travelling at the same velocity in the same

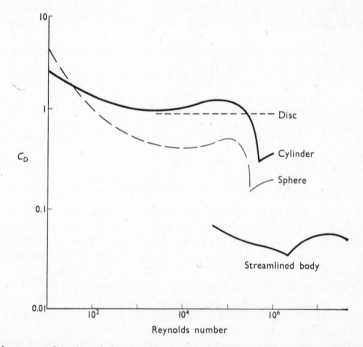

Fig. 4-2 Graphs of drag coefficient against Reynolds number for a disc moving at right-angles to its plane, a long cylinder moving at right angles to its axis, a sphere, and a streamlined body moving along its axis. Reynolds number for the disc has been calculated from its diameter, but for the other bodies the length in the direction of flow has been used in the usual way. (From ALEXANDER (1968))

fluid. These proportions, with length 4.5 times the diameter, are the best from this point of view, at least at the Reynolds numbers (about 10^7) at which this has been investigated experimentally.

The drag coefficients shown in Fig. 4–2 are for bodies with smooth surfaces. Slight roughness has no appreciable effect but excessive roughness increases drag, and the degree of roughness that is tolerable depends on speed. It can be taken as a general rule that roughness only has an appreciable effect at velocity V m/s if the height of the irregularities, from peak to trough, exceeds $1.4/V$ mm in air or $0.1/V$ mm in water.

Many animals which swim or fly have bodies which, after removal of projecting fins and wings, resemble the ideal streamlined form. Streamlining is desirable because the smaller the drag, the less power is needed for locomotion. A few examples are shown in Fig. 4–3. The squid can swim by jet propulsion either forwards or backwards, and backwards swimming

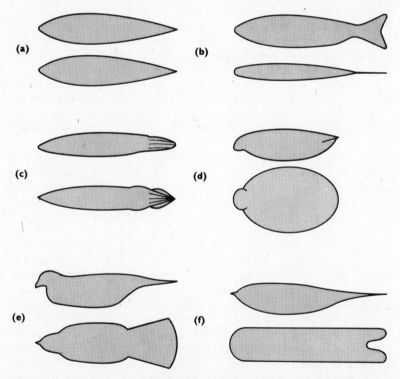

Fig. 4–3   Side and top views of (**a**) a streamlined body of ideal proportions, (**b**) a trout, *Salmo gairdneri*, (**c**) a young squid, *Loligo vulgaris*, (**d**) a water beetle, *Acilius sulcatus*, with the appendages removed, (**e**) a domestic pigeon, *Columba livia*, with the wings removed, and (**f**) a housemartin, *Delichon urbica*, with the wings removed.

is if anything the faster, so it has been drawn facing in the opposite direction to the other animals. It is not certain that the shape which gives least drag for a rigid body is the ideal shape for a fish which bends its body as it swims, but the trout is as near to having a streamlined body with length 4.5 times the diameter as most of the other examples shown.

The drag coefficient of a pigeon body has been determined by measuring the force on a frozen wingless corpse, fastened in a smooth current of air in a wind tunnel (PENNYCUICK, 1968a). The drag coefficients of the bodies of dead fish and of water-beetles (with legs removed) have been determined in the same way with water tunnels (OHLMER, 1964; ROCKSTEIN, 1964–5). The drag coefficients of trout and young squid can also be calculated from measurements taken from cine films of the animals gliding to a halt after a burst of speed (GRAY, 1968; PACKARD, 1969). The coefficients obtained are poor by engineering standards: about 0.4 for the pigeon, 0.3 for the trout, 0.3 for the squid travelling backwards and 0.25 for the beetles. All these are well above the minimum value of about 0.05 for a carefully designed, smooth streamlined body. The drag coefficient of the pigeon is particularly high, but the drag on a pigeon body is nevertheless only about half the drag on a sphere of equal volume. The housemartin body (Fig. 4–3(f)) looks much better streamlined and probably has a lower drag coefficient.

The body of the bonito shown in Plate 2 is an interesting variant of the normal streamlined form, which may be advantageous at Reynolds numbers around 10^7. The deepest part of the body is considerably further back than in Fig. 4–3(a). Bodies of this shape can be expected to have lower drag coefficients in this range of Reynolds numbers than more conventionally shaped streamlined bodies (HERTEL, 1966). This is because drag coefficients of streamlined bodies rise as the boundary layer becomes turbulent at Reynolds numbers above 10^6 (Fig. 4–2). The boundary layer tends to remain laminar as far back as the thickest part, so that less is turbulent, at Reynolds numbers around 10^7, if the thickest part is set well back. The bonito illustrated was about 0.6 m long and spent most of its time in a large aquarium cruising at about 0.9 m/s, when its Reynolds number would be $10^6 \times 0.6 \times 0.9 = 5.4 \times 10^5$, which is too low for this effect to operate. However, its top speed would probably be around 7 m/s, giving a Reynolds number of about 4×10^6, which is probably just high enough. Tunas of similar shape grow much larger and can reach speeds of 20 m/s (GRAY, 1968), which take them well into the range of Reynolds numbers to which their shape could be expected to be best adapted. Dolphins are rather similar in shape and also operate in the right range of Reynolds numbers.

Now we will turn from large swimming animals to small ones. Consider, for instance, *Polytoma* (Fig. 4–4(a)), which is about 15 μm in diameter and swims at about 100 μm/s by beating its flagella. The Reynolds number of its body is then about 1.5×10^{-3}. Water will flow smoothly round it as it swims, as in Fig. 4–1(a). There is no obvious reason why a streamlined

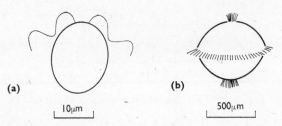

Fig. 4-4 Sketches of (a) *Polytoma*, and (b) the trochophore larva of a polychaete worm.

form should reduce drag, since the purpose of streamlining is to reduce loss of energy to the wake and no wake is formed at such low Reynolds numbers. Little is known about the relative merits of different shapes at very low Reynolds numbers, and this paragraph is intended only as a warning against trying to explain the shapes of small, slow animals in terms which are appropriate only to larger, faster ones. This applies to many small planktonic animals up to about the size of a small copepod. However the water beetle *Acilius*, which is the smallest and slowest of the animals illustrated in Fig. 4-3, swims well within the range of Reynolds numbers in which streamlining is useful. It is a little under 2 cm long and swims at 5–50 cm/s, so its range of Reynolds numbers is about 10^3 to 10^4.

4.3   Flagella and cilia

Flagella and cilia are identical in structure but it is convenient to retain separate terms for them. Those which undulate like swimming eels are conventionally called flagella while those that beat with a more oar-like action are generally called cilia. Only unicellular organisms such as *Polytoma* (Fig. 4-4(a)) rely on flagella for swimming. Various multicellular animals such as rotifers and the planktonic larvae of polychaete worms (Fig. 4-4(b)) and of molluscs and echinoderms, as well as ciliate protozoa, swim by means of cilia. All these animals are small, most of them less than 1 mm long. Is there any essential reason why they should be small?

To answer this question we must know more about drag at Reynolds numbers less than 1. We have seen that it is almost entirely due to viscosity, but we need to know how viscosity is defined. In Fig. 4-5 a flat plate of area A is shown, acted on by a force F which moves it at velocity V parallel to a flat surface. It is separated from the surface by a layer of thickness d of fluid of viscosity η. The fluid at the bottom of this layer is stationary while that at the top is moving with velocity V, so the velocity gradient is V/d. The viscosity is defined by the equation

$$\eta = Fd/AV \tag{4.2}$$

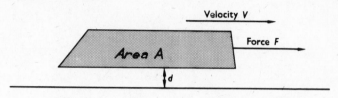

Fig. 4–5   This diagram is explained in the text.

Figure 4–2 shows that the drag coefficients of bodies of various shapes tend to be roughly constant over the range of Reynolds numbers from 10^3 to 10^5. Since the coefficients are calculated from equation 4.1, this means that drag is roughly proportional to the square of the velocity. At lower Reynolds numbers the coefficient is far from constant, and at Reynolds numbers below 1 drag is not proportional to the square of the velocity but to the velocity itself. This is what one might expect from equation 4.2, which shows that the force required to overcome viscosity is proportional to velocity. If the Reynolds number is below 1 the drag D on a body of length l moving with velocity V through a fluid of viscosity η is given by the equation

$$D = k\eta lV \qquad (4.3)$$

where k is a constant whose value depends on the shape of the object.

We can apply this equation to *Polytoma* and other flagellates, which swim in the right range of Reynolds numbers. The drag on the body will be $k\eta lV$, and the power needed to propel it will be (drag × velocity) or $k\eta lV^2$. What power can the flagella produce? They are not passive objects waved from the base, but bend actively all along their length. They are uniform in structure and there is not much difference in diameter between the flagella of different species. The maximum power output of a flagellum is therefore likely to be proportional to its length λ: we will call it $k'\lambda$. At the maximum swimming speed V_{max} of an organism propelled by n flagella each of length λ

$$k\eta lV_{\text{max}}^2 = nk'\lambda$$
$$V_{\text{max}} = \sqrt{nk'\lambda/k\eta l} \qquad (4.4)$$

For a series of flagellates with the same number of flagella and constant λ/l, the maximum speed can be expected to be independent of size. The speeds of flagellates vary between about 20 and 200 μm/s (HOLWILL, 1966) but there seems to be no tendency for speed to increase with size. A large flagellate which could swim no faster than small ones might well be at a disadvantage, and not many flagellates are more than 100 μm (0.1 mm) long. The exceptionally large flagellate *Noctiluca*, which sometimes makes quite a spectacular luminous display in the sea at night, grows to a diameter of more than 1 mm. However, its flagella are rudimentary and it floats passively, unable to swim.

Cilia have the same structure as flagella but are generally far more numerous, so ciliated animals can be expected to have far more power for swimming than flagellates. Certainly ciliate protozoa swim much faster than flagellates, usually at speeds between 0.4 and 2 mm/s.

Cilia propel animals by exerting a backward force, parallel to the surface of the body, on the water near their tips. This means that when a ciliated animal is swimming with velocity V there is a difference of velocity V or more between the water in contact with the surface of the body and at the tips of the cilia. If the cilia have length λ the velocity gradient is V/λ. As no boundary layer would be formed over a non-ciliated body at such low Reynolds numbers this very steep velocity gradient can only be expected to occur over parts of the body which bear cilia, and most of the energy used for propulsion must be devoted to maintaining it. If we assume for the moment that all the energy is so used, and if the area of the animal's surface which bears cilia is A, then, by equation 4.2, the cilia must exert a force $\eta A V/\lambda$ on the water, and the power required is $\eta A V^2/\lambda$. The power available should be proportional to λ and to the number of cilia, which can be expected to be proportional to A. We will take this power to be $k''A\lambda$. Hence at the maximum speed

$$\eta A V_{\max}^2/\lambda \simeq k''A\lambda$$
$$V_{\max} \simeq \lambda\sqrt{k''/\eta} \tag{4.5}$$

The equation shows $V_{\max}$ to be independent of A, implying that an animal with hardly any cilia may swim as fast as one completely covered with cilia (and with far less expenditure of energy since power is proportional to A). This is of course misleading, and is due to our assumption that all the power is used maintaining the steep velocity gradient around the cilia. Gentler gradients which also need power to maintain them must exist over unciliated parts of the body surface, and the total power cannot be less than the value $k\eta l V^2$ which can be calculated from the drag on a passive body (equation 4.3). However, it is clear that propulsion by a band of cilia encircling the body, as on the larva shown in Fig. 4–4(b), must be more economical of energy than propulsion at the same speed by cilia covering the whole surface of the body.

Equation 4.5 indicates that the velocity attainable should be proportional to the length of the animal if λ increases in proportion to body length. This can only be expected to be true up to a certain point. If the total force exerted by the cilia on the water is $\eta A V/\lambda$ and the number of cilia per unit area is constant, the force exerted by a single cilium is proportional to V/λ. The bending moment at the base of the cilium must be proportional to λ times this: that is, to V. (Bending moment was defined on p. 31). There must be an absolute limit to the velocity of ciliary locomotion, dictated by the maximum bending moment which cilia can exert and the maximum number which can be accommodated in unit area of body surface. Large animals will be able to travel faster if they evolve other means of locomotion, which explains why most animals which use cilia for swimming are small.

Much the largest animals which swim by cilia are the ctenophores or comb-jellies. The largest to occur in British waters is *Beroe*, which grows up to 16 cm long, but there are much larger foreign species. The comb plates which they use for swimming each consist of large numbers of cilia attached together: several hundred thousand cilia in 80–100 rows may be present in a single comb plate. Such bundles of cilia must be able to produce ·immensely larger bending moments than single cilia, making much higher speeds theoretically possible. However, ctenophores are very slow for animals of their size.

4.4 Wings

The flight of birds, bats and insects, like that of aeroplanes, depends on the aerodynamic force known as lift. If a streamlined body moves through a fluid in the direction of its axis of symmetry, drag is the only force exerted on it by the fluid. If, however, it is tilted so that its axis is inclined at an angle to the direction of motion the force exerted on it by the fluid has a component known as lift which acts at right-angles to the direction. of motion, in addition to the drag which acts backwards along the direction of motion. Similarly, for bodies of other shapes, there are particular directions in which they can be moved so that only drag acts on them, but movement in other directions causes lift as well. Aerofoils are objects designed to generate lift. The most familiar examples are the wings of aeroplanes and animals. When these are moved through air, tilted at appropriate small angles to the direction of motion, the lift which acts on them greatly exceeds the drag.

Figure 4–6 shows how lift is used in animal flight. (a) Represents gliding. The front edge of the wing is tilted up, relative to the direction of motion, and lift and drag act in the directions shown. If the animal is gliding at a constant velocity in still air, the resultant of the lift and the drag (including the drag on the body) is a vertical force exactly balancing the weight. Figures 4–6(b) and (c) show two mechanisms of flapping flight. The wings move up and down as the animal moves forward so any point on a wing follows a sinuous path through the air, as indicated. In the down-stroke the wing is tilted relative to its direction of motion so that the resultant of the lift and drag on it acts upwards and somewhat forwards, opposing both the weight of the animal and the drag on its body. The upstroke may be merely a passive recovery stroke, with the wing being raised at such an angle that little or no lift acts on it (Fig. 4–6(b)). This is what seems to happen in the normal flight of birds and locusts. Alternatively the wing may be swung forwards in the downstroke and backwards in the upstroke (Fig. 4–6(c)), and be tilted at such an angle in the upstroke that the forces on it act forwards and upwards, helping to support and propel the animal. Flies seem to use their wings in this way, and so do some birds such as pigeons when they are flying slowly (the primary feathers separate in the upstroke, but this is not shown in the Figure).

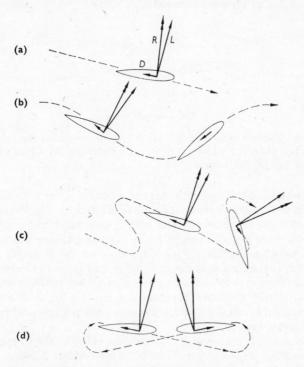

Fig. 4-6 Diagrams showing the paths of wings through the air in **(a)** gliding **(b)** and **(c)** two types of flapping flight, and **(d)** hovering. The wings are shown in vertical longitudinal section at certain points in their paths. The forces of lift (L) and drag (D) and the resultant (R) of lift and drag are indicated.

Lift is essential to the mechanisms of flight shown in Fig. 4–6, but drag is a hindrance. The work which is done when an object is moved against a force is obtained by multiplying the distance moved by the component of the force which acts along the direction of motion. Therefore the wing muscles do work against drag, not against lift. It is impossible to produce lift without drag but the less the drag which acts on the wing when it is producing the necessary lift, the less energy is needed for flight. A good wing should have a shape which makes it possible to obtain sufficient lift without much drag.

Higher ratios of lift to drag can be obtained with long narrow wings than with short broad ones, but excessively long wings would be disadvantageous because their bases would have to be heavy to be strong enough. The ratio of the wing span to the average chord (width of the wing from front to back) is commonly about 7 in birds, bats and insects, though there is a good deal of variation (GREENEWALT, 1962).

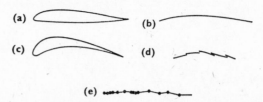

Fig. 4–7 Various wing profiles. (a) A typical streamlined aerofoil. (b) A thin arched plate. (c) A bird's wing near the shoulder. (d) A bird's wing near the wing tip. (e) An insect's wing.

The wing profile (that is, the shape of a section of the wing cut at right angles to the span) also affects the ratio of lift to drag. The largest, and so best, ratios can be obtained either with a wing which has a streamlined profile (more or less as shown in Fig. 4–7(a)) or with one which is a thin arched plate (as in Fig. 4–7(b)), depending on the Reynolds number. Profile (a) suffers less drag than (b) for given lift, and can also provide more lift than (b), at Reynolds numbers above about 10^5. A drastic change in the properties of profile (a) occurs at about this Reynolds number; at lower ones (b) is much the better profile in both respects, and even a flat plate is better than profile (a).

When aerofoils are being considered, the chord is the length used in calculating Reynolds number. Most birds have chords between 2 and 40 cm, and probably fly at speeds between 5 and 30 m/s (PENNYCUICK, 1969). This means that their wings work at Reynolds numbers ranging from about 7×10^3 to 8×10^5. They thus span the region where the properties of streamlined aerofoils change. A thin arched plate would be the best aerofoil for a small bird or for a medium-sized one flying slowly, but a streamlined profile would be better for large birds. The wings of birds are thin arched plates near the wing tips, where the wing consists merely of the overlapping primary feathers (Fig. 4–7(d)). They are streamlined nearer the shoulder where they are thickened by the bones and muscles of the wing, and by the coverts (Fig. 4–7(c)). They are, however, more arched (in technical terms, they have a higher camber) than the wing profiles used on aeroplanes. The wings of bats are more nearly thin arched plates.

The wings of insects work at lower Reynolds numbers. Even for the large fast dragonfly *Aeschna cyanea* Reynolds number cannot be much more than 10^4 (HERTEL, 1966). A streamlined profile would clearly be disadvantageous, and insect wings are in fact thin plates. They are more or less flat in resting insects, but photographs of fruit flies in flight show that their wings are arched for part of the downstroke. Their surfaces are not smooth since they are stiffened by ribs and in some species by slight pleating of the membrane (Fig. 4–7(e)) but these irregularities seem to be too small to have any substantial effect on the performance of the wings at the low velocities involved, at least in most insects.

4.5　How big a bird can fly ?

Small birds such as hummingbirds can hover in still air, but large ones cannot (when kestrels hover they are flying against a headwind at such a speed as to remain stationary relative to the ground). Birds up to the size of a bustard can fly but the largest birds, including the ostrich and the emperor penguin, cannot fly. Is there some physical principle which makes flight in general and hovering in particular more difficult for large animals than for small ones? Is there any reason why flying birds as big as ostriches should not have evolved?

Figure 4–6(d) shows how a hummingbird or insect hovers. The wings beat horizontally. The dorsal surfaces of the wings are uppermost in the forward stroke, and the ventral ones in the backward stroke. In each stroke the wings are held at such an angle as to produce upward lift. The principle is that of the helicopter although the action is a reciprocating one.

Aerofoils, including hummingbird wings and helicopter rotors, produce lift by deflecting air at right angles to their direction of motion. Upward lift is produced by deflecting air downwards. The kinetic energy that must be given to the air can be calculated and in this way it can be shown that the power P_h needed for hovering is given by the equation

$$P_h = (2W^3g^3/\rho\pi s^2)^{\frac{1}{4}} \tag{4.6}$$

where W is the weight of the animal or helicopter, s is the wingspan or the diameter of the helicopter rotor and ρ is the density of air. For isometric birds of different sizes s would be proportional to $W^{0.33}$, so P_h would be proportional to $(W^3/W^{0.67})^{\frac{1}{4}} = W^{1.17}$. If s were proportional instead to $W^{0.5}$, P_h would be proportional to W and birds of all sizes would need the same amount of power per unit body weight, but large birds would have extraordinarily long wings. The hummingbird and the vulture seem both to have typical wing spans for their weight (Fig. 1–5) but if $s/W^{0.5}$ were to be the same for the vulture as for the hummingbird its wing span would have to be 6.6 m instead of the actual value of 2.6 m. Even if such wings were feasible the vulture could not hover, because it could not beat them as fast as a hummingbird beats its wings, so its wing muscles could not produce as much power per unit weight as those of a hummingbird (see p. 6). The wing spans of birds tend in fact to be proportional to $W^{0.39}$ (Fig. 1–5) so according to equation 4.3 the power required for hovering should be proportional to $W^{1.11}$. As size increases the power required must increase faster than the power available so there must be a maximum size above which hovering is impossible.

Hummingbirds can hover continuously for half an hour or more, but larger birds seem unable to hover for more than a few seconds. The largest hummingbirds (about 20 g) may well be the largest birds that can hover without incurring an oxygen debt. A 400 g pigeon can hover (or climb vertically) for about a second, and substantially larger birds cannot hover at all (PENNYCUICK, 1968b). The ability to hover for a moment implies the ability to take off from level ground in still air. Large birds have to take off into the wind, or by diving from a tree or cliff face, or after a taxiing run over land or (like a swan) over water.

The power required for forward flight can be estimated in the same way as the power for hovering, but account has to be taken of the drag on the bird's body. It can be shown that the power depends on the speed and that there is one particular speed at which it has a minimum value P_f. This power which just suffices to keep the bird airborne can be shown from equations given by PENNYCUICK (1969) to be given by

$$P_f = 1.24C_D{}^{0.25}(b/s)^{0.5}P_h \qquad (4.7)$$

where C_D is the drag coefficient of the wingless body, b is the diameter of the body, and s and P_h are the same as before. If $C_D = 0.4$ (the measured value for the pigeon) and $b/s = 0.1$ (a fairly typical value for birds)

$$P_f \simeq P_h/3 \qquad (4.8)$$

This conclusion, that flying needs only one third of the power required for hovering, must be accepted with reservations since the component of power known as profile power has been omitted from both estimates and may not be the same proportion of the whole in both cases (PENNYCUICK, 1968). However, we may reasonably conclude that P_f, like P_h, is likely to be proportional to $W^{1.11}$, and that if a bird were too large it would be unable to fly at all. The maximum weight for forward flight is of course much higher than the maximum weight for hovering.

The Kori bustard, *Ardeotis kori* (Fig. 1-1(b)), weighs about 12 kg and is one of the largest birds which can fly. It seldom flies, and then only for short distances (PENNYCUICK, 1969). Albatrosses and vultures weighing 7–10 kg stay airborne for long periods, but not by active flight. Albatrosses soar in the wind gradients over the sea, and vultures soar in thermals. However, Trumpeter swans (*Cygnus c. buccinator*) may be as heavy as Kori bustards and yet migrate, and so must be able to fly for long periods. They seem to rely on active flight, making no use of soaring, and so are presumably able to fly without incurring an oxygen debt.

Legs 5

5.1 Kinetic energy and legs

Nearly all animals which travel fast on land have legs to run on. The advantages legs give can be understood by considering the kinetic energy changes involved in locomotion, which have already been referred to in the discussion of stepping frequencies (Chapter 1). If a part of the body has mass M kg and is accelerated from rest to a velocity of V m/s, it gains $\frac{1}{2}MV^2$ joules of kinetic energy. The muscles accelerating it must do $\frac{1}{2}MV^2$ joules of work. When it is brought to a halt again energy is used tensing the muscles which bring it to a halt, and the $\frac{1}{2}MV^2$ joules of energy is largely or completely lost (some of it may be saved if the part is moving rapidly backwards and forwards like the human leg in running, since the muscles which stop the leg get stretched elastically and their elastic recoil helps to start it moving again: the leg bounces off its muscles rather as a ball bounces off the ground).

Earthworms crawl without legs but they can only crawl slowly, at about 0.5 cm/s, while a specimen only 2.2 cm long of a fast species of centipede can run on its legs at 0.42 m/s (GRAY, 1968). Each segment of the worm moves in steps, coming to a halt between steps so the whole body is continually being given kinetic energy which is quickly lost. The main part of the body of the centipede moves at a fairly constant velocity and only the legs undergo large fluctuations of velocity. The kinetic energy of the trunk remains fairly constant though kinetic energy has still to be supplied to the legs at every step. Since the legs make up only a small proportion of the mass of the body, the saving of energy is substantial.

The importance of this advantage of legs can be illustrated by estimating the metabolic rate of a worm crawling at the speed of the centipede, 0.42 m/s. Since each segment is at rest for part of the time it would probably have to be accelerated to about 0.8 m/s in every step, so if the worm weighed M kg it would have to be given $\frac{1}{2}M(0.8)^2 = 0.32M$ joules of kinetic energy at every step. A small earthworm might move about 1 cm per step, so 42 steps would be required per second and the power required for the supply of kinetic energy would be $42 \times 0.32M$ watts. If the efficiency of the muscles were 20 per cent (a likely value) and since about 1 cm^3 of oxygen is required for metabolism yielding 20 J the corresponding rate of consumption of oxygen would be $3.2M$ cm^3/s or about 12 cm^3/g body weight hr. This is a high value (Fig. 2–2). Similarly a (wholly imaginary) man-sized animal which crawled earthworm-fashion at 5 m/s, taking 1 m steps, would have to use oxygen at the enormous rate of 200 cm^3/g hr. The use of legs can result in big reductions in metabolic rates needed for locomotion at high speeds.

An ideal animal would run on wheels instead of legs since the kinetic energy of a wheel rotating at constant speed is constant. It is largely for this reason that a man uses less than half as much oxygen when cycling as when running at the same speed (PASSMORE and DURNIN, 1955). However, animals have not evolved wheels and the work used giving kinetic energy to the limbs seems to be around 40 per cent of the work done by the muscles of a running man (CAVAGNA, SAIBENE, and MARGARIA, 1964). It must also be very important in other animals. It can be kept to a minimum if the legs are made as light as possible, and also if their mass is concentrated as much as possible near their upper ends where fluctuations of velocity are least. This is presumably why limbs have evolved with most of their muscles above the joints which they work. For instance, the main knee muscles are in the thigh and not in the calf. The limbs of fast mammals such as deer (Fig. 5–1) are very thick and muscular at the top but very slender at the foot, and the joints of the foot are worked by tendons from muscles considerably further up the leg.

The power needed for running at a given speed depends on the gait as well as on the structure of the legs. When a mammal gallops there are periods when all the feet are off the ground. The longer these are the fewer steps will be needed to cover a given distance, and the less energy will be needed for limb movements. On the other hand, these periods with the feet off the ground are in effect leaps and require energy for raising the centre of gravity. A long leap is necessarily a high one and requires more energy than a short one. A mathematical argument set out by SMITH (1968) leads to the conclusion that the gait requiring least power involves more time off the ground at high speeds than at low ones, and more time off the ground for small animals than for large ones travelling at the same speed. The conclusion seems sound but the argument is probably too simple as it ignores the fluctuations of the kinetic energy of the trunk which may be larger than those of the limbs.

It seems necessary to admit at the end of this section in praise of legs that the serpentine locomotion of snakes involves neither legs nor large fluctuations of kinetic energy. However, since they slide their bodies on the ground they have to do work against friction which is avoided by animals with legs. The fastest speed recorded for a snake seems to be only 7 mph (3 m/s, for a Black mamba, *Dendroaspis polylepis*), but some lizards can run at twice this speed (GRAY, 1968).

5.2 How many legs?

Three legs are enough to support a stool, but a one- or two-legged stool would fall over. For the same reason three is the smallest number of small feet on which an animal can stand firmly. These feet must be appropriately placed: a vertical line through the centre of gravity must pass through the triangle whose corners are the points of contact of the feet with the ground

(Fig. 5-1(a)). If the feet were all to one side so that the line did not pass through the triangle, the animal would topple over. It is of course possible to stand on one or two feet and to walk on two, as birds and people do, but birds and people have large feet and it is easy for them to stand in such a way that a vertical line through the centre of gravity passes through a single foot. A man on stilts is in effect standing on two small feet, and his balance is precarious because it depends on his keeping his centre of gravity very accurately in the right position.

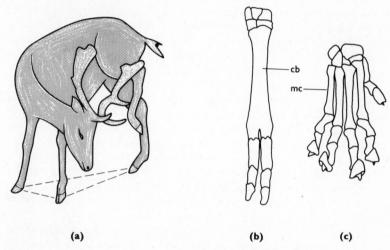

(a)   **(b)**   **(c)**

Fig. 5-1   (a) A Fallow deer, *Dama dama*, standing on three feet. The triangle referred to in the text is shown. (b) and (c) Skeletons of the fore-feet of a deer and a tiger, respectively. The tiger has five toes and five metacarpals like primitive mammals. cb, Cannon bone; mc, metacarpals.

If an animal has four legs, it can move them one at a time and so always have three feet on the ground as it walks. If it moves its feet in the right order (and four-footed animals seem always to move their feet in this order when they walk) it can keep the vertical line through its centre of gravity always within the triangle formed by the supporting feet (GRAY, 1968). Every position it adopts as it walks is stable. This stability is obtained at the expense of slowness, and it is more usual for each foot to be raised a little before the previous one is set down, so that periods of instability with only two feet on the ground intervene between the periods of stability. In faster gaits amphibians, reptiles and mammals all tend to move their limbs in pairs, so the body rests on two or fewer limbs and is unstable for at least most of the time. When the limbs move in pairs they are off the ground for as long as they are on it and do not have to move any faster (relative to the body) in the forward stroke than in the backward stroke.

Insects have six legs. When walking slowly they sometimes move only one or two legs at a time so that there are always four or five on the ground. When they move fast they tend to move three limbs at once (these are the front and hind legs of one side and the middle leg of the other). Each foot is on the ground for only half the time so the forward stroke is no faster than the backward one, but there are always three feet on the ground and the body is always stable. Only a few insects adopt a fast gait involving unstable phases with fewer than three feet on the ground.

Four legs is the minimum number for stability in slow walking, but it is not enough for stability in fast gaits in which each foot is off the ground for half the time. For this, six is the minimum. The most successful walking animals are the relatively large four-legged amphibians, reptiles and mammals, and the relatively small six-legged insects. Can the difference in number of legs be explained by the difference in size?

Consider how far an animal is likely to be thrown off course by a gust of wind when it has fewer than three feet on the ground. For the sake of simplicity we will base our calculations on a wholly imaginary animal whose body is a cylinder of length l m and radius r m, and whose legs are so thin that we can ignore both their mass and the drag which acts on them. The frontal area exposed to a wind which strikes the body sideways on is $2rl$ m^2, the density of air is 1.3 kg/m^3 and the drag coefficient of a cylinder in the range of Reynolds numbers which concern us is about 1 (Fig. 4–2). Hence by equation 4.1 the drag exerted on the body by a side wind of velocity V m/s is $\frac{1}{2} \times 1.3 \times 2rlV^2 = 1.3rlV^2$ Newtons. The volume of the body is $\pi r^2 l$ m^3 and if its density is about the same as that of water (1000 kg/m^3) its mass will be about $1000\pi r^2 l$ kg. By Newton's second Law of Motion the drag will give this mass an acceleration $1.3V^2/1000\pi r = 4 \times 10^{-4}V^2/r$ m/s^2. An acceleration a continuing for time t results in displacement $\frac{1}{2}at^2$, so if the acceleration we are considering lasts for t s the body will be displaced $2 \times 10^{-4}V^2t^2/r$ m. If the part of the cycle of limb movements in which two or fewer feet are on the ground lasts for t s, the animal may be blown this distance off course before another foot which can correct the balance is placed on the ground. Is this distance likely to be substantial? We will find out, giving V the reasonably low value of 5 m/s which makes the distance $5 \times 10^{-3}t^2/r$ m.

Consider first a medium-sized insect, with $r = 0.002$ m (2 mm). When it ran fast it would probably make about 20 cycles of leg movements per second (GRAY, 1968) and if it had only four legs a phase with two on the ground might last half a cycle, giving $t = 0.025$ s. The displacement would be $5 \times 10^{-3}(0.025)^2/0.002$ m or about 1·6 mm, a substantial displacement for a small insect. Now consider a mammal of about the size of a dog, with $r = 0.1$ m. It might make 3 cycles of limb movements per second, giving $t = 0.17$ s. These figures give a displacement of only 1.4 mm, which would be insignificant for an animal of this size. The comparison is not entirely fair because wind velocities decrease as one approaches the ground. This decrease is so variable that it cannot easily be allowed for, but would not be so sharp as to invalidate the conclusion that the insect would be blown off balance far more easily than the mammal. It may be for this reason that natural selection has favoured six legs for insects but not for the generally larger terrestrial vertebrates, allowing insects but not vertebrates to have three feet always on the ground even in fast gaits.

Four and six have been explained as the minimum numbers of legs for constant stability in slow walking and for fast gaits, respectively. Is there an advantage in keeping the number of limbs low? If a horizontal force F is to be applied to the ground by a vertical limb the minimum cross-sectional area of skeleton required at a given height above the ground is, by equation **3.3**, $kF^{\frac{3}{2}}$, where k is a constant. If two limbs are to be used, each exerting $F/2$, each will require a cross-sectional area $k(F/2)^{\frac{3}{2}}$ and the total of their cross-sectional areas will be $2k(F/2)^{\frac{3}{2}} = 1.26kF^{\frac{3}{2}}$. Thus increasing the number of limbs makes necessary an increase in the total weight of the limbs, and it is advantageous to have as few limbs as possible. For the same reason, a foot with only one or two toes can be made lighter than one with more toes. This is why the number of toes has been reduced to one in horses and two in deer etc., and why the five metacarpals or metatarsals of their ancestors have been reduced to a single cannon bone in both (Fig. 5-1(b)).

5.3 Posture and stability

Insects and crustaceans stand with their feet well apart on either side of the body while mammals stand with their feet close together under the body (Plate 3). Having the feet well apart will make an animal less liable to be overturned by wind or (if it lives in water) water movements. Is there any reason why this posture should be more necessary for arthropods than for mammals?

Consider again the simplified animal of Section **5.2**, whose body is a cylinder of length l m and radius r m. Let it stand with its feet so placed that lines from the points where they touch the ground to the axis of the cylinder make an angle θ with the horizontal (Fig. 5-2). What velocity of wind, striking it sideways, will overturn it?

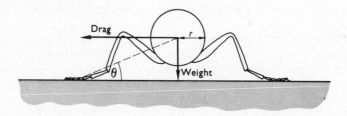

Fig. 5-2   This diagram is explained in the text.

We have to consider the two forces shown in the Figure. The drag exerted by a wind of velocity V m/s has already been estimated as $1.3rlV^2$ N. The mass of the body has been estimated as $1000\pi r^2 l$ kg so the weight, expressed as a force, is $1000\pi r^2 lg$ N. At the critical wind velocity V_c m/s at which the upwind feet are on the point of being lifted off the ground and the animal is on the point of overturning, the moments about the downwind feet are balanced and

$$1000\pi r^2 lg \cos\theta = 1.3rlV_c^2 \sin\theta$$

$$V_c = (2400 rg \cot\theta)^{\frac{1}{2}} \qquad (5.1)$$

For a dog, $r\simeq 0.1$ m and $\theta\simeq 80°$ (Plate 3) giving $V_c\simeq 20$ m/s. For a blowfly, $r\simeq 0.002$ m and $\theta\simeq 30°$ giving $V_c\simeq 9$ m/s. However, if the fly stood like a dog with $\theta=80°$, V_c would be only 3 m/s and it would be overturned by a very gentle breeze indeed. It is significant that the Pauropoda which stand higher on their legs than most arthropods (θ seems to be about 50°, GRAY, 1968) live in leaf litter and in crevices in the soil where the air must generally be still. A mammal can of course lean into the wind to avoid being overturned, and many insects have adhesive pads on their feet which could be used to fasten down their upwind feet and help to prevent overturning. However, it is clear that a low value of θ can give an advantage to a small animal but is unlikely to be helpful to a large one.

Now consider a crustacean which walks under water. Crayfish live in rivers in which there are, of course, currents. Lobsters, crabs etc. live largely in shallow water near the edge of the sea, where there are tidal currents and the water movements of waves. Do they need to spread their legs well apart to keep themselves stable? The currents to which they are exposed are not in general as fast as the winds experienced on land, but water is about 800 times as dense as air and so, by equation 4.1, a water current gives about as much drag as a wind $\sqrt{800}=28$ times as fast (not necessarily exactly as much, since the drag coefficient depends on the Reynolds number). Many crustaceans are likely to experience more drag than terrestrial animals of similar size. The vertical force which resists overturning is the weight in water which is seldom more than 20 per cent of the weight in air, even for thick-shelled crabs and lobsters. Well-spread legs and low values of θ seem essential for crustaceans which walk under water.

Not only do insect and crustacean legs differ in number and position from mammal legs, but they have a different type of skeleton. They and other arthropods have an external skeleton covering the body, while mammals and other vertebrates have internal skeletons. This difference has already been discussed in a book in this series (CURREY, 1970) in which reasons are suggested why an external skeleton is best for small animals and an internal one for large animals, and especially for large land animals.

5.4 Legs and weight

When an animal stands stresses are set up in the bones and muscles which support its weight. Stress is force per unit area. The cross-sectional areas of corresponding bones and muscles in isometric animals would be proportional to (body weight)$^{0.67}$, so the stresses due to body weight would be proportional to (body weight)$^{0.33}$. However, large animals are made of the same materials as related small ones, and the stresses they can stand are the same. A very large terrestrial animal must therefore have

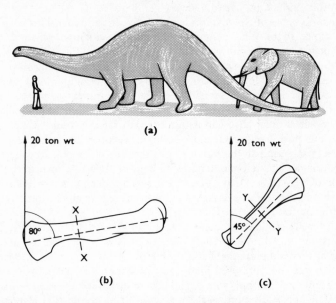

Fig. 5-3 (a) An *Apatosaurus* as it probably appeared in life, with a man and a large African elephant drawn to the same scale. (b) The femur and (c) the tibia and fibula of *Apatosaurus* showing forces which would just break them, according to the rough calculations described in the text.

disproportionately thick limb bones and muscles, or must stand and walk in such a way as to minimize stresses, or both. The problem of support must set an upper limit to the size of animals which walk on dry land, and the question whether any terrestrial animals have ever reached this limit will be discussed. The problem does not arise in aquatic animals whose weight is supported by buoyancy. The largest whales weigh over 130 tons and are considerably larger than any terrestrial animals that have ever lived.

A large African elephant weighs only about 7 tons, but some of the extinct sauropod dinosaurs were very much larger. These cannot of course be weighed in the flesh since we know them only as fossils, but their weights can be estimated by making scale models of them as they probably appeared in life, measuring the volume of each model and calculating from this and the likely density of the dinosaur the weight of the living animal. The largest known dinosaur, *Brachiosaurus*, probably weighed about 80 tons and *Apatosaurus* (Fig. 5–3) about 30 tons. It has been suggested that these very large dinosaurs could not support themselves on dry land, but spent their lives wading with much of their weight supported by buoyancy. This could possibly be true of *Brachiosaurus* but there is fairly convincing evidence from fossil footprints that *Apatosaurus* or a similar sauropod could walk on land (COLBERT, 1962).

The limb bones of *Apatosaurus* were massive (their dimensions are given by GILMORE, 1936). What margin of strength can we suppose them to have had, above the minimum needed to carry the animal's weight? It seems from the shape of the animal that the major part of the weight must have fallen on the hind limbs, and it can reasonably be supposed that they must have carried about 20 tons. When the animal took a step this weight must have been carried by one hind limb. A quick calculation indicates that the bones must have had ample strength for this while vertical, but how about positions (which must occur in walking) when they slope? A vertical force acting on the end of a bone then has a component which acts at right-angles to the bone, tending to bend it. It was shown in the discussion of trees (p. 31) that a given force sets up much greater stresses in a slender cylinder if it acts at right-angles to the axis of the cylinder than if it acts along it.

The tensile strength of bone is about 10^8 N/m² (1 ton wt/cm²). This information, and the dimensions of *Apatosaurus* bones, can be used to estimate very roughly the bending forces the bones could withstand. Equation **3.2** could be used for this if the bones were circular in cross-section but they are not, so a slightly more elaborate equation has to be used. The bones are solid, so there is no need for the further elaboration that would be required if they had marrow cavities. It seems from the shapes of the femur and of the tibia and fibula that they would be most likely to break at XX, YY (Fig. 5–3(b) and (c)). The equation indicates that they would just be broken at these places by forces of 20 tons weight acting as indicated in the Figure. If the muscles were strong enough, the animal could have taken a step with the tibia and fibula of the supporting hind limb at up to about 45° to the vertical, and could have done so with the femur almost horizontal. The force exerted by a foot while on the ground would probably not be absolutely constant, and would probably rise briefly above 20 tons weight (especially if the dinosaur stumbled and had to save itself). However, the bones seem amply strong enough for walking so long as the animal did not take excessively long steps. The footprints show 2.4 m steps, which though long seem reasonably modest for an animal with hind legs more than 3 m long. It is possible to imagine a larger animal which had relatively thicker limb bones, or which shuffled along with smaller steps (so keeping its limb bones more vertical) or which lived more dangerously with less spare strength in its limb bones. It is not at all clear that a larger terrestrial animal would be impossible. However, it seems certain that their size made large dinosaurs much less agile than small reptiles, just as elephants are less agile than small mammals. Some small animals can exert on the ground forces many times their own weight: it can be calculated from their jumping ability (see ALEXANDER, 1968) that a 0.1 kg bushbaby can exert at least 1.5 kg weight on the ground, and a 0.5 mg flea over 50 mg weight.

Modern reptiles walk and stand with their feet well out on either side of the trunk and their humeri and femora roughly horizontal. This gives them, perhaps unnecessarily, high stability of the type discussed in Section 5.3. If large dinosaurs had stood like this they might have needed even more powerful limbs and would certainly have wasted a great deal of energy maintaining tension in their limb muscles. The power required for this is presumably proportional to the force the muscles must exert (which is proportional for a given posture to the weight W of the animal) multiplied by the length of muscle exerting the force (proportional in isometric animals to $W^{0.33}$). If so, the power required for standing by isometric animals in a given posture is proportional to $W^{1.33}$. Since metabolic rates tend to be proportional to $W^{0.75}$ (Fig. 2–1), standing with the legs splayed out to either side is more taxing for a large animal than for a small one. The footprints show that large dinosaurs stood like mammals, with the feet directly below the trunk.

5.5 Walking on water, and on the ceiling

Many insects walk and run on the surface of water. Some, such as the pond-skaters (Gerridae), spend most of their time on the surface and others, such as adult mosquitoes, only alight briefly on it. All are fairly small. Is small size essential for walking on water?

Insects are supported on water surfaces by surface tension acting on their feet. Figure 5–4 shows how the foot rests on the water and how the

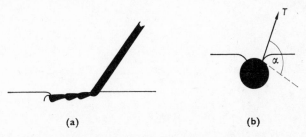

(a)   (b)

Fig. 5–4 Diagrams showing (a) in side view and (b) in section the foot of an insect standing on the surface of water.

surface is deformed by it. The contact angle (α in Fig. 5–4(b)) of water with insect cuticle is high, so if the foot is pushed well down into the water as shown surface tension acts more or less vertically on it. The surface tension of pure water is about 70 mN/m: that is to say, a force of 70 mN acts at an edge, 1 m long, of a water surface.

A large pond-skater might weigh 50 mg, and so need a force of 0·5 mN to support it. This could be provided by surface tension on its feet if the total perimeter of the feet was at least 0.5/70 m = 7 mm. Six feet each 1 mm long (and thus having a perimeter of just over 2 mm each) would be more than adequate. Not only can pond-skaters rest on water, but they can go on doing so when the surface tension is reduced by detergents until it falls to about 40 mN/m (BRINKHURST, 1960).

A small woman might weigh 50 kg, a million times as much as a pond-skater. To stand on the surface of water she would need feet with a total perimeter of 7 million mm: that is, 7 km. No large animal could conceivably evolve feet big enough to support it by surface tension on water. This is of course because the force exerted on feet by surface tension is proportional to their length while the weight the feet have to support is proportional (in isometric animals) to (length)^3.

Similar considerations help to explain why large animals cannot walk upside down on the ceiling like flies. The largest animals which can achieve this feat are the smallish lizards known as geckoes, and it is uncertain how they do it. They have on their feet large pads bearing microscopic bristles, and it may be that the ends of these bristles fit so closely to the irregularities of the ceiling as to adhere by van der Waals forces. The ends of the bristles are shaped like tiny suckers and it has been suggested that geckoes adhere by suction, but this possibility has been eliminated by showing that the feet can stay fixed in a vacuum. Whatever the mechanism, it seems likely that the force it can produce will be proportional to the area of the soles of the feet, which in isometric animals would be proportional to (length)^2. The force it must withstand is the weight of the animal, proportional to (length)^3. It is therefore more difficult for large animals than for small ones to walk on the ceiling.

Size Limits 6

Much of this booklet is filled by discussions of the appropriateness of various shapes to organisms of different sizes, and of size limits for organisms with particular features. It has, for instance, been shown that flattening is necessary for a large flatworm but not for a small one (Chapter 2) and that well-spread legs give valuable stability to a fly, but are unnecessary for a dog and would be very exhausting (even if possible) for a sauropod dinosaur (Chapter 5). It has been shown that neither a very small warm-blooded animal (Chapter 2) nor a very large animal propelled by cilia (Chapter 4) is feasible. However, the absolute limits of size for organisms have not been considered.

The lower limit of size is probably set by the complexity of life. Viruses do not possess all the equipment needed for life, but depend on the living processes of the cells they inhabit. If they are excluded the smallest known organisms (and the smallest known cells) are mycoplasms about 0.3 μm in diameter (PIRIE, 1973). It has been claimed that there are even smaller mycoplasms about 0.1 μm in diameter capable of independent life, but this does not seem credible. If they existed and had a normal cell membrane (about 8 nm thick) this membrane would occupy nearly half the volume of the cell. Most of the other half would be filled by the ribosomes needed to synthesize 45 enzymes, which has been estimated to be the minimum needed for life. There would be no room for the rest of the apparatus of life.

There is no obvious single factor which sets an upper limit to the size of organisms, though there are as we have seen factors which limit the sizes of organisms of particular types. The largest of all organisms are *Sequoia* and *Eucalyptus* trees a little more than 100 m high. Their height may be limited by the problems involved in raising sap, but this topic has already been discussed in this series of booklets (SUTCLIFFE, 1968) and will not be discussed here.

References

Items marked * are suggested for further reading. The others are sources of data used in this booklet.

*ALEXANDER, R. MCN. (1968). *Animal Mechanics*. Sidgwick & Jackson, London.

BAILEY, N. J. T. (1959). *Statistical Methods in Biology*. English Universities Press, London.

BENEDICT, F. G. (1938). *Vital Energetics*. Carnegie Institution, Washington.

BERGER, M., HART, J. S., and ROY, O. Z. (1970). Respiration, oxygen consumption and heart rate in some birds during rest and flight. *Z. vergl. Physiol.*, **66**, 201–14.

BRINKHURST, R. O. (1960). Studies on functional morphology of *Gerris najas* Degeer (Hem. Het. Gerridae). *Proc. zool. Soc. Lond.*, **133**, 531–559.

BROWN, M. E. (edit.) (1957). *The Physiology of Fishes* (2 vols). Academic Press, New York.

CAVAGNA, G. A., SAIBENE, F. P., and MARGARIA, R. (1964). Mechanical work in running. *J. appl. Physiol.*, **19**, 249–256.

*COLBERT, E. H. (1962). *Dinosaurs*. Hutchinson, London.

*CORNER, E. J. H. (1964). *The Life of Plants*. Weidenfeld & Nicolson, London.

*CURREY, J. (1970). *Animal Skeletons*. Arnold, London.

FRASER, A. I. (1962). Wind tunnel studies of the forces acting on the crowns of small trees. *Rep. Forest Res.*, **1962**, 178–83.

GATES, D. M. (1965). *Energy Exchange in the Biosphere*. Harper & Row, New York.

GILMORE, C. W. (1936). Osteology of *Apatosaurus* with special reference to specimens in the Carnegie Museum. *Mem. Carnegie Mus.*, **11**, 175–300.

*GOULD, S. J. (1966). Allometry and size in ontogeny and phylogeny. *Biol. Rev.*, **41**, 587–640.

*GRAY, J. (1968). *Animal Locomotion*. Weidenfeld & Nicolson, London.

GREENEWALT, C. H. (1962). Dimensional relationships for flying animals. *Smithson. misc. Collns*, **144**, (2), 1–46.

HEGLAND, N. C., TAYLOR, C. R. and MCMAHON, T. A. (1974). Scaling stride frequency and gait to animal size: mice to horses. *Science, N.Y.*, **186**, 1112–3.

*HERTEL, H. (1966). *Structure, Form, Movement*. Reinhold, New York.

*HILL, A. V. (1950).The dimensions of animals and their muscular dynamics. *Science Progr.*, **38**, 209–30.

HOLWILL, M. E. J. (1966). Physical aspects of flagellar movement. *Physiol. Rev.*, **46**, 696–785.

KLAAUW, C. J. VAN DER (1948). Size and position of the functional components of the skull. *Arch. néerl. Zool.*, **9**, 1–176.

KROGH, A. and WEIS-FOGH, T. (1951). The respiratory exchange of the desert locust (*Schistocerca gregaria*) before, during and after flight. *J. exp. Biol.*, **28**, 344–57.

LAVERACK, M. S. (1963). *The Physiology of Earthworms*. Macmillan, New York.

MAGNUSON, J. J. and PRESCOTT, J. H. (1966). Courtship, locomotion, feeding, and miscellaneous behaviour of Pacific bonito (*Sarda chiliensis*). *Anim. Behav.*, **14**, 54–67.

NICOL, J. A. C. (1967). *The Biology of Marine Animals*, 2nd edn., Pitman, London.

OHLMER, W. (1964). Untersuchungen über die Beziehungen zwischen Körperform und Bewegungsmedium bei Fischen aus stehenden Binnengewässern. *Zool. Jb.* (*Anat.*), **81**, 151–240.

PACKARD, A. (1969). Jet propulsion and the giant fibre response of *Loligo*. *Nature, Lond.*, **221**, 875–77.

PASSMORE, R. and DURNIN, J. V. G. (1955). Human energy expenditure. *Physiol. Rev.*, **35**, 801–40

PEARSON, O. P. (1947). The rate of metabolism of some small mammals. *Ecology*, **28**, 127–145.

PEARSON, O. P. (1948). Metabolism of small mammals with remarks on the lower limit of mammalian size. *Science*, **108**, 44–46.

PENNYCUICK, C. J. (1968a). A wind-tunnel study of gliding flight in the pigeon *Columba livia*. *J. exp. Biol.*, **49**, 509–26.

PENNYCUICK, C. J. (1968b). Power requirements for horizontal flight in the pigeon *Columba livia*. *J. exp. Biol.* **49**, 527–55.

*PENNYCUICK, C. J. (1969). The mechanics of bird migration. *Ibis*, **111**, 525–556.

*PIRIE, N. W. (1973). On being the right size. *Ann. Rev. Microbiol.*, **27**, 119–32.

RIEDL, R. (1964). Die Erscheinungen der Wasserbewegung und ihre Wirkung auf Sedentarier im mediterranen Felslitoral. *Helgolander wiss. Meeresunters.*, **10**, 155–86.

ROCKSTEIN, M. (edit.) (1964–5). *The Physiology of Insecta* (3 vols.). Academic Press, New York.

SCHMIDT-NIELSEN, K. (1972). *How Animals Work.* Cambridge University Press.

SCHOLANDER, P. F. (1955). Evolution of climatic adaptation in homeotherms. *Evolution*, **9**, 15–26.

*SMITH, J. M. (1968). *Mathematical Ideas in Biology*. Cambridge University Press.

*SUTCLIFFE, J. (1968). *Plants and Water*. Arnold, London.

*THOMPSON, D'A. W. (1942). *Growth and Form*. Cambridge University Press.

VOGEL, S. (1970). Convective cooling at low airspeeds and the shape of broad leaves. *J. exp. Bot.*, **21**, 91–101.

WAINWRIGHT, S. A. and DILLON, J. R. (1969). On the orientation of sea fans (genus *Gorgonia*). *Biol. Bull.*, **136**, 130–39.

WEIS-FOGH, T. (1964). Diffusion in insect wing muscle, the most active tissue known. *J. exp. Biol.*, **41**, 229–256.

WHITNEY, R. J. (1942). The relation of animal size to oxygen consumption in some fresh-water turbellarian worms. *J. exp. Biol.*, **19**, 168–75.

WHITTAKER, R. H. and WOODWELL, G. M. (1968). Dimensional and production relations of trees in the Brookhaven forest, New York. *J. Ecol.*, **56**, 1–25.